PLACES of the SOUL

. . . an inspiration to all those who care about the influence of the environment on Man's health and well-being.

Barrie May, *The Scientific and Medical Network*

At last an architect has written a sensitive and caring book on the effects of buildings on all our lives.

Here's Health

This gentle book offers a route out of the nightmare of so much callous modern construction. I was inspired.

Colin Amery, *The Financial Times*

Christopher Day

Places of the Soul

Architecture and Environmental Design as a Healing Art

Thorsons
An Imprint of HarperCollins*Publishers*

Thorsons
An Imprint of HarperCollins*Publishers*
77–85 Fulham Palace Road,
Hammersmith, London W6 8JB

Published by The Aquarian Press 1990, 1993
This edition by Thorsons 1999
9 11 13 15 17 16 14 12 10 8

Christopher Day asserts the moral right to
be identified as the author of this work

A catalogue record for this book
is available from the British Library

ISBN 1 85538 305 5

Printed in Great Britain by
Woolnough Bookbinding, Irthlingborough,
Northants

Contents

1

Architecture: Does It Matter?

Architects tend to think architecture matters. Not everyone else does. To many people buildings are expensive, for it's what goes on *inside* them that matters.

The argument continues that it is better to have a good teacher (or craftsman, mother, designer, manager, etc.) in an ugly shed, barrack, pre-fab, tower-block flat, etc. than a poor one in a beautiful room. But few of us are exceptionally good or exceptionally hopeless; we are middling and need support. So, how supportive is the barrack to a middling teacher? Ultimately, how good is the teaching?

Children, even those below the age at which they express any aesthetic interest in their surroundings, behave differently in different environments. Even mature adults feel, think and act differently in different surroundings; though their actions may be under more conscious control, their world outlook, sensitivies and thought mobility are influenced. I sometimes wonder what sort of qualities and sensitivities my work would have if I worked in a different office — perhaps a harsh, rectangular, smooth-

Whether you like this or not, this is not architecture. It is a photograph of a building. A semantic distinction? On the contrary. One is a static view, chosen by someone else, freezing a transient moment of light, season, weather, approach, life. . . the other is, influences, or is an interrelating part of, our total physical surroundings: like the photograph, its effects extend beyond the physical to touch our feelings. While, however, the photograph may focus our attention yet we may scarcely notice our surroundings, one we can turn away from, the other is the frame in which we live. We don't just look at

architecture, we live in it.

This book is illustrated with photographs. They are incomplete and inadequate fragments of experience, for architecture is for much more than the eyes. It is for life. And that is why it is such a powerful tool — often devastating, but potentially health-giving.

Photographs are selective. Most people's interest is in the people, whereas architects tend to concentrate on buildings — often without any hint of occupancy. While many of these photographs show empty rooms to avoid intrusion, try to imagine them in use for their specific functions.

surfaced, evenly lit, glossy one such as many architects work in.

Many people believe that artistic ability is a matter of inborn genius, but I am convinced that the main factor is *commitment*. Likewise, aesthetics is much less a function of money than of care, but care costs time. In a world where time means money, the less care put into buildings — in design, construction and use — the cheaper they will be, but as few people want cheap-looking buildings, deceptive appearance, from brick facings to cardboard structured doors, chipboard furniture to glossy fronts and cut-price rears, has become commonplace.

We are rapidly building a world where deceptive appearance inadequately screens the primacy of profit over care. It is the world in which most people spend much of their daily lives and in which children grow up and learn — from their surroundings as well as from people — the values that will support them in later life.

A lot of people complain about modern architecture. They complain about *performance* aspects of old buildings (such as dampness) but about *environmental* aspects of new ones (such as anonymity). Other than architects, few people *think* about architecture, but many people *feel* it. It is those who don't that I feel sad for, for their aesthetic feelings have been blunted, even obliterated — and architecture must carry much of the blame.

We all know that 'other people' tend to be negative, critical and opinionated; often identifying things unfairly and condemning them unjustly. It was an eye-opener to me to experience positive feelings from unexpected people when, about 1973, I built a house. All sorts of people, passing by, asked to look and commented in terms of real appreciation. They were farmers, carpenters, factory-workers, postmen and the like — many of whom lived in, wished to live in, or built, bungalows. I realized that many people *choose* those sort of buildings because they are the only choice they can imagine.

These blinkers of imagination shape and are shaped by the speculative building industry and other vested-interest manipulators of wants such as manufacturers of kitchen fitments. Architectural fashion on the other hand is guided by what is individualistically new, and this tendency is intensified by the focus of architectural magazines on buildings as dramatic (and usually unpeopled) objects. These buildings *are not experienced in that way by the people who use them*. Magazines appear to have a greater influence on architectural students than do their teachers, good or bad. It is likely that they also influence at least some practising architects. Yet the consciousness that they tend to support — *building* consciousness (and with strong 'image' characteristics at that) — has nothing to do with creating *places for people*.

Small wonder that architecture is sick! It can make people feel ill and *be* ill, and we can — in theory — measure the biological consequences (although in fact our knowledge is inadequate at this stage). To make matters worse, not everybody gets ill from breathing radon, fomaldehyde or mould spores; there is only a tendency. When we get to aesthetic qualities, these are widely considered to be subjective, a luxury that can be applied later after the practical problems have been solved and *if* we can afford it. I take absolutely the opposite view.

In good health, I have taken my son to hospital clinics and felt only half alive after sitting for hours in rectangular grid-patterned, vinyl-smelling, fluorescently-lit, over-heated corridors. The brutal vandalism of buildings unfeelingly imposed upon the landscape can have the same effect. Architecture can be life-suppressing or even crushing, not only to our finer sensitivities but to our feelings of freedom. In some places one feels a trapped statistic, not a valued member of society; in others the buildings tower over one as though with menace. Often, however, what architecture does to people is much more subtle, and recently the social and psychological effects of barren, harsh, unwelcoming architecture have become more and more the subject of popular concern.

Over half a century ago, Rudolf Steiner remarked that there is 'as much lying and crime in the world as there is lack of art'. He went on to say that if people could be surrounded by living architectural forms and spaces these tendencies would die out. When first I heard this I thought, what bourgeois nonsense! After all, the roots of crime are complex, socio-economic underprivilege playing a large part. If, however, we broaden our definition to include exploitive abuse of people and environment, and

recognize that we are talking not about inevitable destinies but about tendencies, it is easier to see what he meant. Animals invaryingly respond to environmental stimuli whereas humans have the ability to transcend the situation. To rise above the level of automatic reaction requires, however, that we consciously direct our lives. None of us is perfect in this respect and that is why in any statistical sample, while some individuals may not do so, many people tend to react to stimuli in predictable ways.

In this century, the world has seen rapid urbanization, frequently associated with the disruption of habitual social supports to accepted moral codes. There are endless social aspects of the new environments people meet, but if we just consider the architectural surroundings the predominant tendency has been to produce forms, spaces, shapes, lines, colours and relationships between elements (not to mention air, electric fields, noise, and so on) that are life-sapping, dead in quality.

In the absence of aesthetic nourishment, the emotional part of the human being is left to seek fulfilment by indulgence in desires. Deceptive appearance, especially of building finishes, are quite normal, so are sterile spaces which depend upon deception — cosmetic surface, mood-manipulative lighting, for instance — or upon contents to make them habitable. Surrounded by harsh hardness, the aesthetic sensitivities, and with them moral discernments, are blunted. Surrounded, as most of us are most of the time, by lifeless man-made mat-

It is no wonder that places like this have become notorious for their crime rate. The issue is less that of easy opportunity, but of faceless, depersonalized, uncaring, insensitive harshness.

ter, it is no wonder that the attitude of trying to get what you can get has grown so strong that it is even enshrined politically.

Indissolubly intertwined with other causes and effects, these are the consequences of the world we have built around us. Yet architecture, although built of matter, need not be dead: it *can* be life-filled. Its constituent elements and relationships can sing — and the human heart can resonate with them.

Most people, myself included — but possibly architects excepted — don't normally look at our surroundings. We breathe them in. We look at picture postcards or at views from viewing platforms, and these can be interesting. However, the experience only touches our hearts when it becomes an ambience we can breathe; most of the time we don't notice our surroundings and then they can work upon us without any conscious resistance on our part. In fact we spend so much time in or near buildings that it is true to say that most of our environmental experience is affected by architecture.

Architecture is potentially a dangerous tool. Environment can be used to manipulate people: because we so readily take our surroundings for granted and rarely bring them to full consciousness, they can be used to influence our actions. We don't need to look back to Nazi stadiums with their theatrical mood-distortion devices. When we go into a boutique with a 'vibrant world' mood created by music, textures, colours, levels and diagonals as background but with light and layout focus on goods we are free to touch, desires are being sharpened; their satisfaction seems linked with purchasing.

Mood enhancement becomes manipulation when pressure is brought to bear. Few supermarkets have much in the way of an *inviting* mood, but with lighting, sign and display colours and background music, they can subtly enhance the excitement of buying. Compare how many shelves of goods or focused display lights in your local supermarket are brightly lit and in the warm, bright, active colours or sparkling white and how many are softly lit and in the blue range.

Some of this comes from studied psychological technique, some from inspired display design, and some is perhaps quite devoid of conscious intent. Not only in shops, not only to make money or power out of other people, but in every aspect of environmental design, we must recognize that whatever we do affects the human being, the surroundings, the spirit of places and the wider world. It has human, social, biological and ecological implications. We only need to live briefly in a different environment to recognize how much our surroundings have formed us and our society in sensitivities, in values, in way of life.

Architecture is such a powerful agent that how it is worked with matters a great deal. A great deal indeed! How it affects people and places, how design and construction can be approached to bring health rather than illness, is the subject of this book. There are many ways to go about this but to describe things one has not personally experienced risks abstraction, wishful thinking and second-hand opinion. While I

prefer therefore to describe how I myself go about doing things, I am anxious not to preclude ways which might suit other people better. Examples are, by nature, local; the issues that underly the process by which they come into being are universal. Different people, in different locations, will need to evolve different solutions.

We each of us start life differently and live through a completely personal stream of experiences. One person's style therefore can never honestly suit another's. In this way style is very personal; while many can recognize and perhaps appreci-

Without consciously looking at them, we breathe in our surroundings with all our senses. In some places, the outer, communal, world only makes us feel exhausted and unwell. No wonder some people seek inner, private, relief by artificial stimulants.

ate it as an intellectual symbol for an outlook on the world, unless that style can be transcended their feelings remain untouched.

I try not to have a style, but it is easy to lapse into one. What I hold as my inspiration is a way of looking at things to gain insight into what they really are and do, so that appropriate forms can arise. This is relevant to all people in all places. My subject is the built environment, my examples localized in time and place, but the issues are equally applicable in England or New England, in urban Tokyo or suburban Sydney, the townships of South Africa or the forests of Scandinavia. Any building, any place, in any type of land- or townscape, in any culture, climate or country has effects such as I describe. Wherever it may be located, and however differently it may outwardly appear, if architecture is to be health-giving it must work with these themes.

2

Architecture With Health-Giving Intent

Architecture is but a part of the built environment. Inside a building parts of the building become the whole environment; outside it forms only part of our surroundings. We rarely experience larger buildings as architectural objects, but where we do, it is usually because they are forceful and dominating. Such buildings impose their presence on us and — most particularly — are imposed upon their surroundings. They also lend themselves better to photography and so are favourite subjects for architectural magazines.

Wherever I go, I see buildings imposed. Imposed because they are inappropriate, insensitive. They are crystallized monologues which do not meet the needs of people or place.

In the days of hand-power it was easier to go round a tree-root or a boulder or follow a contour than go straight through. The lines that resulted — for path, field boundary or building placement were, for pragmatic reasons if no other, in conversation with the landscape. Powerful machinery finds it easier

to disregard the irregularities of the surroundings. When you get to know old buildings and old fields you can start to notice how the climate differs when you step beyond their boundaries. This sort of sensitivity in placing does not occur when you design things on paper. Paper design and mechanical construction have changed the relationship of buildings to surroundings much more dramatically than first appears.

Nowadays you can design a building in one country to be built in another. No longer is it a case of famous European architects designing buildings for America; now all over the world buildings designed out of one culture are placed in another. Such buildings typically have artificial indoor climate control and are reached by car. They can therefore be sited anywhere in the world, but they belong nowhere. Other than a few examples of these, the buildings I show photographs of are not transplantable.

There is considerable danger in transposing ideas from one culture,

Above *Often our experience of buildings is not as freestanding objects but of boundaries of space. The quality of this boundary is a major ingredient of the quality the place will have. Whether in the country or town, boundaries made of unrelieved straight lines are so harsh and lifeless. If qualities of movement, life, harmony, gesture and resolution of dynamic forces can be given to lines, shapes, forms and spaces they can bring an influence of life to the place a building bounds. These qualities, normally found in curves, can also be achieved with hand-drawn straight lines in conversation with each other. Imagine for a moment this ridge and eaves to be single dead-straight lines.*

Below *From a distance smaller buildings are experienced as objects in relation to their surroundings. Spacial relationships between buildings give the first hint that a place — not just building-objects — exists.*

one landscape, to another. North American concepts of freedom do not sit easily in Nicaragua. Buildings appropriate in Surrey do not fit in Midlothian. One family of materials, and therefore language of form, may suit buildings in Wales; elsewhere, with different materials, climate, culture and all the other contextural considerations, a quite different language of form is needed. The underlying issues of what environment does to people are not limited to national, regional or parish boundaries but, if they are to be appropriate, the forms they give rise to will be intensely local. I find this attunement to the local situation, where all sorts of automatic assumptions have to be cons-

When the relationship between building and surroundings is such that the building can be seen as an object, it has a responsibility not to offend its surroundings. In an agricultural setting a balance between the two elements seems appropriate. In wilder surroundings the building needs to be more reticent.

ciously re-examined, to be amongst the hardest aspects of the design process whenever I work in a new locality, and I challenge any architect to dispute this.

Before we even consider how architecture can be health-giving, we have to make sure that our ideas and our buildings are not imposed! The international style — be it functionalism, post-modernism or any other -ism — and the inter-regional styles — such as bungalows — disregard what is appropriate for particular places and the people of those places. Neo-vernacular and revivalist reactions have something hollow about them: they also seek to impose a singular idea, in this case plucked from a particular period of past history. All such approaches are more concerned with *style* than *responsiveness*.

Architecture has such profound effects on the human being, on place, on human consciousness, and ultimately on the world, that it is far too important to bother with stylistic means of appealing to

PLACES OF THE SOUL

fashion. It can have such powerful negative effects that we must also think, can it, if consciously worked with, have equally strong positive effects?

Anything with such powerful effects has *responsibilities* — power, if not checked by responsibility, is a dangerous thing! Architecture has responsibilities to minimize pollution and ecological damage, responsibilities to minimize adverse biological effects on occupants, responsibilities to be sensitive to and act harmoniously in the surroundings, responsibilities to the human individualities who will come in contact with the building, responsibilities not only in the visual aesthetic sphere and through the outer senses but also to the intangible but perceptible 'spirit of place'.

These responsibilities involve energy conservation ranging from insulation, organizing buildings around a focal heat source like a heating/cooking stove, conservatory or hypercaust wall, re-use of waste heat — for instance retrieving heat from waste water or refrigeration coils — to alternate energy production such as solar heating. They involve careful selection of building materials and the ways they are put together, both with regard to occupants' and manufacturing and building workers' health, and to the environmental impact of products from primary extraction to demolished rubbish. Such wider criteria cast a new light on, for example, tropical hardwoods, the extraction of which in almost all cases entails massive ecological damage; on plastics, which commonly require some 15 or so syn-

thesis operations, each around 50% efficient, so that the final product is only some 0.002% of the original material;[1] and on water, new reservoirs for which are brutally destructive of upland communities.

This attitude of responsibility involves putting away stylistic or individualistically-enhancing preferences in favour of listening to what the place, the moment, and the community ask for.

This design process can be continued right through construction. This depends on hand construction. This may sound unrealistically out of date, but the fact is that mechanical systems do not have sufficient flexibility; the growing building cannot be adapted when potential benefits or short-comings become apparent. Hand construction also gives textural scale, for instance with bricks, slates or wood instead of large mechanically erected panels. Where opportunities exist for the builders to become artistically involved in their work, such buildings have a distinct soul even before they are occupied. The spirit of place can develop because of, not in spite of, the building. The method of construction and form of contract therefore have a bearing on the spirit of a building quite apart from its appearance!

Thinking about users means thinking of buildings as spaces, their outsides as boundaries to spaces. Small rural buildings we may experience as objects *in relationship with their surroundings*. Large urban buildings are more commonly experienced only as the

[1] Holger König, *Wege zum Gesunden Bauen*, Ökobuch (Freiburg), 1989, p.33.

boundaries of space. Space is to live in. Objects are frozen thoughts. The one is life-enhancing, the other, if big enough, is threatening, dominating, stealing sunlight with its huge shadows or tricking our sense of orientation with its reflections. I remember as a student how much time I wasted drawing carefully composed elevations and how little I spent on sections, internal perspectives or views of relationships with surroundings. Now it's the other way around! Now I am less interested in objects but in *places.*

The place cannot speak in human words but we can listen to what it asks for, what it will accept. When I see places where the charm is in part due to the buildings, I realize this is the standard I must aim for!

Everything we build is new, but once it has been there a year or two it can look out of place and out of time. It can look old — only to reveal its deceit on closer inspection — or it can be neither old nor new, but eternal and inevitable. I try to design places which are not in conflict with the stream of the past — everything that has contributed to the present — but are not old fashioned or imitative: places which are inspired out of the future — the world of ideas, ideals, inspiration and imagination — but still have their feet in the reality of the present moment — for the present, however inspired by the future is *built upon the past.* Ignoring the stream of the past is vandalistic; concentrating too strongly on it risks meaningless preservation or revivalism.

Towns without a past tend to have social problems. It can take several generations until they stabi-

It is sometimes hard to imagine that a place could be as attractive and inevitable without the buildings.

lize. Conserved historic places are little better: they can be claustrophobic to live in, and falsely cosmetic. Past and future need each other: the past informs, the future inspires. At the meeting point is the informed, inspired present — the point at which deeds are born.

To a large extent all of this is about stopping architecture being harmful — those places which, for instance, make one cringe or feel depressed or ill in. Sometimes we blame the noise, the air conditioning, the fluorescent lights, the crowds, the proportions, the smell — but all of it comes down to architecture, whether the circulation of people, acoustics, out-gassing toxins, colours, spatial aesthetics or construction detailing.

The trouble is we become dulled to these things. We don't notice the

17

To be harmonious, the new needs to be an organic development of what is already there, not an imposed alien.

noise, the bad air, the harsh conflict of hard-edged shapes and forms. We become immune to the negative forces in our environment — *and that is when they do us most harm!* Our sensitivities and our senses become dulled and our language and unconscious approach to daily life begin to reflect our surroundings. City people speak more sharply than country people — at the extreme, if we put their voices in pictures, one is a confusion of sharp, hard edges and the other a mush of sleep-inducing rounded forms. And yet modern cities are only a few generations old. Like speech, social sensitivities are also hardened by harshness and ugliness in the surroundings — ask anyone who works with adolescents.

Who, looking at his own ex-perience can deny that architecture at its best or worst extremes speaks a strong language? Mass housing is quite different from homes that are individually and lovingly made in every detail — one is provided for *statistics*, the other for *individuals*. It makes a lot of difference whether things are designed *for* people or *together with* them. Architects hope their buildings will last for several generations, so however much they design with occupants in mind, they will never meet all of them. But unless I can design something nourishing to *my* soul — *nourishing*, not just nice, dramatic, photogenic, novel — I can't expect it to be nourishing to anyone else.

We tend to think first of visual aesthetics. Concern with visual aesthetics is the major part of most architects' work, mine included. We all know that a picture is worth a thousand words, that the optic nerve is massive compared to that from other sense organs — but a smell can take us back to forgotten childhood memories, music can take us into another world . . .

All the senses have their parts to play — in ugliness or in beauty — but all too often they are considered in isolation. When together, giving the same message, they start to speak of the underlying essence of a place. When sensory messages conflict, environmental improvements are just playing with cosmetics. Just as *Concorde* may look like a beautiful bird but doesn't sound like one, a beautiful well-landscaped architectural façade fronting a heavy main road is a nonsense. All you are aware of is the bombardment of noise. It's as hollow and meaningless as synthetic fresh bread smell outside a fast-food restaurant. The fashion for polyurethane lacquered wooden furniture comes from 'visual only' consciousness. When you touch it, the wood is hard, shiny, cold and does not breathe. It doesn't smell of wood and it looks glossy — a surface, not a depth of colour.

Television shows us a world in sight and sound only. It is a most deceptive and deprived picture of reality for we hardly notice the absence of other sensory information. The senses — all together — give a picture of a reality which is not adequately described by any *one* sense, a reality which we call spirit, the spirit of a person, event or place. More than just the appearance or comfort, it is this spirit which affects us deeply.

To be healing, a place must be harmonious, bringing change as an organic development so that new buildings seem not to be imposed aliens but inevitably belong where they are. They must respond to the surroundings and be responsible, seeking to minimize pollution caused by their materials. But places — and buildings — must be more than that: they must be nourishing to the human being.

The concept of health presupposes sickness. We all know what is is like to feel ill, but *why* we become ill is not so readily understood. Science has shown how illness can be triggered by material agents such as viruses, bacteria — triggered, not caused. Not everyone catches an epidemic. The ideal germ warfare agent — one which will infect 100 per cent of a population — has yet to be found. In some areas up to 50 per cent of autopsies show tuberculosis scars on lungs of people who never became ill with what used to be regarded as an incurable disease.[2] Some diseases show quite different symptoms in different people.

Symptoms express and release what is going on within the body — a high temperature for instance shows the struggle between antibodies and pathogens. In the same way illness expresses and releases inner and less visible disharmonies. To understand sickness and healing, whether medically or architecturally, we need to understand something of the different levels of the human being.

We all know the human body is a physical lump — so much flesh, bones, blood, etc. Knowledge of

[2] Benjamin Spock, *Baby and Child Care*, New English Library, 1975.

this physical body is essential if we want to be able to reach shelves, sit comfortably, avoid back problems, and so on. Ergonomics and space allocation for specific positions and activities are taken for granted, and like every other architect's office I have a book of anthropometric data which shows me what space I need for a moving skeleton.

What distinguishes the human being from the corpse is the fact that it is alive. Architecture can either support or damage physical health, supporting it, for instance, by keeping the body within an appropriate tempered environment — neither too hot nor too cold, too bright nor too dark.

Different kinds of heating and lighting feel healthy or unhealthy, inviting or unpleasant. The light from a log fire has a similar spectrum to sunlight. Its radiant heat seems particularly warming — to soul as well as body. It may be inefficient from the energy point of view but many people like it — you can feel well in front of a fire. Many people complain of a dry throat, stuffy nose and lethargy with forced air central heating; some feel claustrophobic and oppressed by it. The physical causes are incorrect ionization and humidity of the air (which also aggravate, and are aggravated by, any problems of indoor air pollution), airborne dust (especially that cooked by the heater) and even, undifferentiated temperature, with overheated air and underheated radiant surface. With such different effects on our well-being, it comes as no surprise that what *feels* better *is* better.

The sparkling quiver of candlelight, however inadequate in brightness, has a life that the

mechanically even vibration of electric (especially fluorescent) light just does not have. So also does the daylight in a room lit by several windows, creating an interplay of lights, hues and shadows from different sky directions. Mono-directional light from a single source, be it window or window-wall, does not have this life.

It is no accident that these 'feel alive', for they are life-enhancing, in a strictly biological sense: growth and other hormones have been found to be controlled by the pituitary, pineal and hypothalamus glands, and these are stimulated by light. Not any light, but gentle rhythmical living light, particularly daylight with its many moods and colours from different directions endlessly changing throughout the day. That is why light from two windows in two walls from two sky

colours is always more pleasant and *healthy* than one. Just as smell is nature's way of telling us that things are good or bad for us so there is a meaningful coincidence between the aesthetically satisfying and the physically healthy. What nourishes the soul nourishes the body.

The science of building biology is still in its infancy and many of its assertions are challenged, particularly by industries whose products are threatened. But even in the absence of scientific data we can to some extent *feel* when a place is healthy and physiologically life-supporting and when it is not. We share this realm of life and biological effect with everything that lives — but we are more than that. However insensitive we may be, environment does have an effect on our feeling life. Tourism (and the picture postcard business) depends upon places that people choose to visit, if only to *look* at them.

There are places, like widenings in a corridor with a window seat, that induce casual social meetings, and places, like lifts, that stifle such interplay. Similarly, there are shapes, like round tables, which bring people into community, and others, like uninterrupted corridors or long rooms, which do not. A narrow, low, not quite straight, invitingly

Light from two windows on different walls gives a life to the light which can even be seen in the frozen moment of the photograph. This life in the light is as necessary for biological as for psychological health, the pituitary gland and the soul both being nurtured by living light and both deprived by dead light. The physiological and aesthetic effects are inseparable.

If you want to institutionalize a building, you need corridors. If you want to raise movement from A to B to become a renewing, preparatory experience, you can use a cloister. Cloisters are semi-outside spaces, around a garden; if glazed, they cease to be cloisters. None the less, we can build some of their quality into passageways so that any future destination can take second place to the experience of where you are now. How else can eternity live in every moment?

textured and lit corridor for unhurried uses, like that of a monastery cloister, can be a real delight; a smoothly surfaced, evenly lit, straight corridor for large numbers of people in a hurry is quite the opposite, and it makes even the most well-meaning building into an institution.

Architectural psychology studies the environmental requirements of places in which we can feel good,

private, sociable, and so on. On the whole, however, we don't need to look at a book to know what effect design decisions will have — but rather we can refer to our own experience, using ourselves as an instrument of assessment. Of course, everybody has different preferences and associations so here one must try to distinguish between what are individual or cultural and what are universal responses. Black, for instance, we associate with death — but in the East they use white! Colours however have physical characteristics and physiological effects from which no one, whatever their personal likes and dislikes, is immune. Geometry has similar universal effects, as has proportion as it is founded upon the measurements of the human body; so do scale and speed.

Once we recognize that many qualitative aspects of environment have universal effects in addition to personal and cultural ones, we must recognize that the human being — each of us — is potentially an *objective instrument of assessment*. That which many dismiss as 'subjective' can in fact be assessed objectively: entirely new distinctions between objective and subjective can thus arise based upon these new criteria.

What makes the human being really *human*, however, is the ability to distinguish what would be the right or wrong way to act. Unlike animals we must go beyond instinct, habit or behaviour-conditioned learning and use our thinking and moral and aesthetic sensitivities to consciously choose our actions. In our surroundings we also make distinctions as to what we like or dislike. We can be nourished by artistic qualities which go beyond mere psychological technique. To uplift the spirit, places need to be in some way artistic.

With this approach we can develop a qualitative vocabulary to nourish the human soul, but to be healing we must go further. If we picture the human as a being of four levels as I have described — a being of body, life, feelings and moral individuality — we can see disharmony at the most inward level expressed in progressively greater substance as it is transformed through each level until it becomes a physical aberration, like a tumour, which can even remain on a dead corpse. Treatment, by surgical, chemical or other means can destroy physical and psychological ailments at those (outer) levels, but unless the deep-seated disharmonies are addressed new ailments have the habit of emerging. Healing means transformation at the inmost level, and the individual can only do this for himself. How then can this be accomplished? What has it got to do with architecture?

It is hard even to recognize the need for such inner transformations, and harder still to start them. Something from outside, such as counselling, homoeopathic medicine, or some other agent is needed to initiate and support the process. Environment is one such agent: it can provide nourishment, support and balance for the human spirit as much as it can starve, oppress and pervert it. The more it works with universal rather than personal qualities, the more it can transform feeling responses from personally indulgent desires to artistic experiences.

But environment — even static, mineral, architectural environment — does more than this. Our environment is part of our biography. It is part of the stream of events and surroundings that help make us what we are. In Churchill's words, 'We shape our environment and our environment shapes us'. If, for a brief tolerant moment, you entertain the possibility of reincarnation

The sequence of preparatory experiences we pass through to approach, enter and use a building do more than affect our experience of it. They change our inner state which can both enhance our receptiveness to health-giving qualities in our surroundings, and trigger transformative processes in our inmost being. All healing is founded on such inner transformations, albeit initiated by outer agents. Threshold, sequence and 'oasis' have therefore important health-giving functions.

and destiny, you are faced with the question: why have we chosen one particular life path, one progression of environments, and not another?

There are punitive and positive theories of reincarnation. The latter suggests that in our path of personal development, we need to meet and resolve those things previously unresolved or side-stepped. Throughout each life we draw to us opportunities, often in the outward form of obstacles, that we need. This is not to say that our surroundings should necessarily provide a wide range of obstacles; rather, if it can provide qualities which have been meaningful in earlier lives, the resolve to transform obstacles into opportunities in *this* life can be strengthened.

Timeless qualities have a profundity that can bring us to a threshold

The experience of walking along this path is woven of alternating obstructed and expansive views, steps and turns and, especially, the textures of light and shade. It gives a particular sense of coming down to the lower town or up to the town centre. Like the surrounding architecture the path is pleasant though unexceptional but the journey is a delight.

experience of inner change necessary to set in motion the healing transformation of the inner self. Entering into the experience of a work of art brings us to such an inner threshold, and this is the foundation of art therapy.

Our surroundings are potentially the most powerful artform we experience in our lives. Whether they will bring illness or healing depends upon all of us whose decisions and actions shape human environment.

3

Architecture as Art

When you try to observe what the innate essences of things are and how and why such things affect us, it is easy to see that there are rules that underlie all universal, and therefore profound, experiences. To be healing, however, we have to transcend thresholds, we have to move from rules to art. But what is art?

I have heard poetry described as that which makes your skin prickle when you read it. That is close to my definition of art — the experience of something which leaves you never to be the same again. It has brought an inner step forward. Medical, psychological and spiritual healing involve processes by which something outer is brought to the patient so that he or she can make an inner step. It is a process of enabling, not of manipulating, just as healing is quite distinct from medical, psychiatric or ideological 'treatment'.

The arts — whether painting, architecture, even cooking or gardening — are involved with raising material *matter*. In this sense art is the imbuing of matter with spirit,

and it is this spirit that the user unconsciously experiences and that has a healing influence. But how to make this step from rules to art?

Unfortunately, it doesn't seem enough just to have good intentions or theoretical understanding. Good intention remains abstract so long as it is not worked out in deeds and products; graceless actions bring a disharmony which works to negate those good intentions, whereas artistic work roots them more fittingly in matter. When I left architecture school I saw many of my fellow students whose ideals I admired abandon them as untenable in the 'real world'. What needless tragedy, for, if true, good ideals — however unfashionable — are essentially practical, craving to be worked out practically, artistically, in the world.

It is necessary to *cultivate* a sense for beauty, for the artistic. I say 'necessary' because our culture tends to suppress this sense, and 'cultivate' because everyone has it latent within them. It used to be so strong that pre-industrial common

25

people could not make a spoon, a cart, a boat, even a house look ugly. To do so would have been like a crime against themselves. Everything, from reaping the corn to blessing the meal or carving a chair, was an action giving thanks for God's creation, an artistically satisfying activity. All they made and did was essentially functional: there was no time, energy or space to make anything without a practical purpose; beauty and utility were inseparable.

Today we find the reverse. Beauty and utility are widely regarded as completely separate streams: we all need utility, but beauty is considered to be an indulgence, peripheral to our main concerns in life. We have the means now to produce quantity — unnecessary quantity — and quality is a secondary consideration. In the eighteenth century audiences used to weep during concerts; today not uncommonly the emotions are *compelled* by decibel power.

We have become used to the idea that money may be spent to beautify places for recreation and leisure but that places of work or for practical activity should be shaped first and foremost by utilitarian considerations. The implication is that if half one's working life is spent as efficiently but inartistically as possible, the other half is free to be artistic and inefficient. The assumption that practicality and aesthetics are mutually contradictory only holds good when the definitions are narrowed: practical to the monetary, aesthetic to indulgent self-expression. Yet when we realize the relationship that exists between

aesthetics and health, this severance of utility and beauty can be seen to be as unhealthy as it is philistinic.

Nowadays it is quite impossible to return to pre-industrial values, for these were quite unconscious and habitual. Their forms were dominated by stereotypes, their inner and outer horizons confined. Today, thank goodness, if we choose to consider beauty and utility as inseparable we do so in full, committed consciousness. We can consciously choose the direction of our artistic work to be appropriate to the needs of the circumstance rather than personally indulgent. But what do people need from the architectural environment?

All of us from time to time find ourselves in states of mind like boredom, insecurity, loneliness or stress which need something outside ourselves to provide a balance. Where the environment can offer interest, activity and intriguing ambiguity, timeless durability and a sense of roots (in place, past and future) in the wider natural world with its renewing rhythms, sociable places and relaxing atmospheres for the socially shy, and harmony, tranquillity and quiet soothing spaciousness, it can provide support as the first step to recovery. Where it cannot provide for these soul needs we find the classic modern phenomenon — dependence. Dependence on prescribed or narcotic drugs, alcohol, television, consumerism. We can find endless 'soul needs' to suit our ever-changing moods, but there is a more limited range that *must* be found in our surroundings if they are to be supportive. How many homes have a social — and physical — warmth focus of a hearth? How many people have access to the freshness of happily singing water?

When an understanding of the universal characteristics of our artistic vocabulary is added to a sense for beauty and an attunement to the needs of the soul, the results are both artistic and appropriate. Spiritual functionalism we could call it. For instance, colour can be used functionally — a Steiner kindergarten room should support imitative and imaginative activity within a warm, supportively secure, almost dreamy environment. The appropriate colour lies in the warm pink range. On the other hand a classroom for older children needs an environment which helps the

The classic example of the underlying archetypal idea responding to the locally individual influences of environment, active through both place and time. Architecture works with the same polarities.

teenagers bring the outer activity of earlier childhood more into themselves inwardly; this encourages the intellect to be more active in contrast to the earlier education. The appropriate colour lies in the cool blue range — but *very* delicate. But *how* the rooms are coloured, the exact hues and shades and variations, depends upon a dialogue between colour, natural light and space. A kindergarten in Oslo or Milan would be quite different, even if the practical needs of colour are more or less identical.

This conversation between the universal and the particular, the inspiration and the local circumstances — the moment — is the same as between the archetypal oak tree principle and the battering winds focused by topography. The results are both individual and universal in the same moment. It is the same as between the principles of cosmic geometry and the

demands of climate, site surroundings and so on, or between pure idea and the demands of building materials and construction — for instance gravitational or tensile principles in a stone arch or a tent.

Conventionally there are two streams of architecture — high and low. One concerns itself with cosmic rules — proportion, geometry and classically-differentiated elements representing universal principles: relation to the earth, to the vault of the heavens, to the vertical boundaries of free-stretching space, as experienced in the human limbs, head and torso; also to the finely tuned shape of space, form, and so on. Like classical music it must work within but rise above these rules to become art, something to elevate the spirit of man. This is the stream of great architecture — temples, cathedrals, sometimes palaces and civic buildings. In scale and commitment to a singular idea such buildings often dominate the surroundings.

The other stream is the vernacular. Its keynote is response to climate, materials, social form and tradition. It concerns itself very much with textures, meetings of materials, and tends to be rich for the senses. Almost without fail, the resulting landscape and townscape forms warm the soul. Internally, however, the stereotyped idea of how one should live from generation to generation often tends to be too strong.

The high architecture stream is inspired by cosmic ideas, the vernacular stream is rooted in daily reality — one is learnt by prolonged esoteric study, the other by making,

doing and building, by mud, dirt and wood shavings. Both are artistic but neither is complete or balanced without the other: they need to be brought into conversation.

Real conversation is never a compromise. Something dies in a compromise, but in a conversation something new is born. It is in this something, this 'spirit of conversation', that the universal and the uniquely particular are fused into a work of art. As with design conversations with clients, what arises is better and *more appropriate* than either of us could have done on our own. Appropriate is a key word. Things are only appropriate if they meet the needs of the circumstance — and there are many needs: the surroundings, the wider community, the health of the earth, all have needs as well as those of the building users.

Very few of these needs can be voiced in words. We have to listen to the unspoken, listen with all our senses. It is this listening as an exercise that develops our sense of what is *right* — our sense of beauty. If we look at the world around us the places which are most rich in life are meeting places, and not only cafés and city squares. In nature, life is at its most vigorous where the elements meet — in the warm sun-drenched marshes, the humid jungle. When we seek rejuvenation in natural surroundings we are drawn to those places where the spirit of place is strongest — where there are meetings between elements — places which emphasize the meeting of, for instance, earth and sky or water and rock.

If we sit and watch these meetings — how at the rock the water swirls, eddies, splashes, gurgles, sings, smells wet and cool — we realize that what is happening is too rich and mobile to be depicted or described. Also, if we can immerse ourselves into the mood of what is going on, we can become in a way attuned to 'rock-wateriness', to the spirit of what is happening.

To cultivate abilities to work with the *appropriate* and the *beautiful*, we need to do exercises like this. There are exercises we can do on our own, like looking every day at the dawn which is different each day and changes every moment, yet at every moment has something eternal, as do the endless but ever-changing waves of the sea.

These are things we can do on our own and there are many other exercises to suit other individuals — but what they all have in common is that they are *listening* exercises. Listening to what is already there is the first step in any meaningful architecture just as it is the first step in any therapy. The physician 'listens' to his patient — to what he says, how he speaks, his appearance, face, and so on. Listening is the fundamental requirement for any conversation — or indeed for any healthy social process! We only go forward by recognizing that which the process enables to come into being that wasn't there before.

Success depends upon putting personal preferences aside and *listening* without any judgement (except as to truth) — even listening to the unpalatable. In architecture this means listening to the needs of people — which they rarely can voice properly, to the needs of the place, to the opportunities which are concealed and which become more and more apparent as design, then construction, then use, progresses. Techniques and procedures such as I describe later can make this easier, but the vital element is the cultivation of the ability to listen.

Architectural demands so often lead in different directions, in potential conflict — like energy conservation and biological effects, the straight and the curved, cosmic geometry and organic response to environmental circumstance — that the results will be one-sided and disastrous unless they can be brought into a conversational balance. Similarly, architectural elements need to be brought into conversation or they fight against each other.

I am not just talking about what is nice or not but about what is nourishing for the human spirit. To be nourishing, things must match what we need, just as a stoker and a meditating hermit need different diets. Our surroundings therefore must satisfy necessary material functions; they must provide the right biological climate; they must give support to our life of mood and feeling. But to carry architecture beyond the threshold of the materially useful, the biologically supportive or the emotionally satisfying, we need to cultivate and bring together both the inspiration which gives moral force to our ideas and the sense of listening to environment which makes those ideas appropriate. This interweaving conversation between idea and material can only exist in the artistic sphere.

4

Building for Physical Health

It is easy to see how much harm human activities do to the world. We can look at almost every product as bought at the price of environmental or human damage. Out of this attitude rising to (albeit limited) popular consciousness around 1970 grew the 'restricted damage' approach to building. Zero-energy, ecologically autonomous houses and self-sufficient farming became a select fashion. Two decades later it is easier to take a wider overview and see that architecture, like any other art-form, can bring spiritual benefits to humanity and to the earth which outweighs the material damage that it causes. The world would be a poorer place without Chartres Cathedral, but it took a lot of stone quarrying. We *can* build wholly biodegradable buildings from earth, straw and small branches, but *all* buildings which satisfy the performance criteria we expect in the developed world cause ecological damage to some extent. Their materials are almost entirely mined from our surroundings — even

modern forestry is mining. Many cause pollution or use considerable energy in their manufacture. In use, buildings consume a lot of energy, the production of which has ecological consequences.

So what is this damage? What price do we pay? We can read in the newspapers of poisoned groundwater, radioactivity in food, dying forests and seas and collapsing ecology in whole regions. This isn't just what other people do. It is product and by-product of the way we build and live. This sort of pollution hardly existed before the industrial revolution and much of the worst of it has only been invented in the last few decades. It isn't the only way to live and to build, but it is the normal way these days. To make changes therefore that are acceptable to the people who will pay for them and that perform acceptably is not easy. But if we are aiming to build architecture which has a health-giving influence, we need responsible foundations.

There are several major aspects of building which affect the environment, including toxins entering the biosphere as the result of industrial processes — often off site. Britain produces five million tons a year.[1] If you have ever been to a chemical waste dump you will know the chill dead horror, the dying cry of the earth.

Energy in one form or another always has an environmental price. Even hydro-electric power, though less dangerous to the world as a whole than nuclear power, has a high local price in ecological disruption. Greenhouse effect global warming, now irreversible, is a consequence of profligate energy consumption. Much more energy is used in running buildings than in constructing them and manufacturing their materials. Running buildings consumes some 50% of UK energy, a proportion little changed over two decades as rising expectations of comfort have paralleled improvements in thermal insulation.

In addition to dispersed effects consequent on their industrial support base, buildings themselves — their materials, location, services and design — have local effects. They affect the health of people as well as places.

These simple issues have complex implications, not least because if considered uni-dimensionally they are often in conflict with each other: excessive concern for energy-saving in the last two decades has been a major cause of 'sick' buildings, those which cause health problems for their occupants. Draught-proof-ing has led to buildings being less well ventilated than ever before in human history — yet we need fresh air to live! Almost all industrially-produced thermal insulations have significantly harmful characteristics: sharp mineral fibres, dusts, gaseous emissions, even radioactivity. 'Non-industrial' products exist, but the supply of cork, coco-fibre insulation or, if you don't insulate, firewood cannot be drastically increased without damage to trees. Locally traditional materials can cause health problems such as rheumatism and bronchitis associated with dampness and mould spores.

Conservation of energy on the other hand is simpler to deal with. Only a few decades ago thermal insulation was practically unheard of. Now the regulations have reached a standard that I no longer feel necessary to double!

It makes energy sense to take account of local climate. In my own area the problems are mostly those of wind-cooling. Ground-hugging buildings do well here. Heating buildings means heating space, so the smaller and more compact the heated volume the less energy needed — a point I have not always been able to convince energy-conscious clients on! To minimize energy use as well as other effects on the surroundings, like relative scale and shadow size, the first step is to think small. Compact spacial arrangement does not necessarily mean cramped environment. As I will discuss later, other qualities can have a greater effect than dimensions.

From the energy point of view alone, different climates require

[1] *Environment Digest* No 10, March 1988.

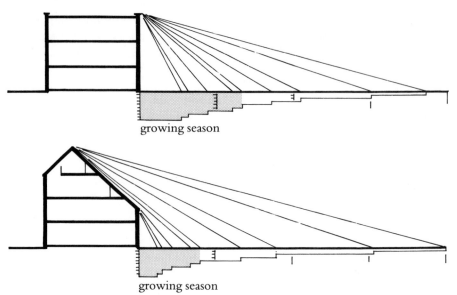

growing season

growing season

different building forms. Hot climates need shaded, airy spaces, such as verandas or court-yards. Different humidity, temperature and night /day variations have led to different traditional materials and natural cooling layouts which give rise to vernacular forms ranging from large ventilated roof-spaces (which can double as crop-drying lofts) to high thermal capacity mud-brick buildings.

Traditional building forms are not to copy; copies are so dead. We need however to be sure that we have carefully thought about what previous generations took for granted before we depart too radically from them. If not, these buildings will depend upon large energy inputs for heating or cooling — and energy has an even higher ecological price than it does a monetary one.

We can laugh at British colonial administrative buildings standardized in design with corrugated

Shape affects size of the shadow cast. Shadow may benefit car parks, but not gardens, parks and pavements. Fewer plants grow in shadow and they grow less well. In areas of permanent shadow very little grows at all — beneath the lank vegetation is bare mud, poor to play on. The consequent low level of soil life is slower to break down organic refuse such as bird droppings, dog mess and old leaves. Less plant growth means less air-cleansing. Shadows bring gloom and poor health to cities — they are a product of size, orientation and shape of buildings.

iron roofs and fireplaces even in the tropics (I am told that the fireplace was cool enough to stand in, elsewhere was like an oven). On the other hand we take it for granted that office blocks are mechanically cooled and ventilated. Yet whenever buildings require energy inputs to provide a physical environment that could have been achieved by design means, we must recognize that building design is responsible for completely needless pollution.

It is now well established that it

is cheaper to conserve than to produce energy. Many people think first in terms of alternative energy gadgetry. I think of these last. None the less some alternative energy is simple to produce. The economics depend on your accountancy assumptions: solar hot water can be proved to be a money-saver *and* never to pay for itself. My question to clients therefore is, do you think it a right thing to have or are you only looking to save money?

There is nothing complicated about solar water heating. I normally use a system in which aluminium fins clip to copper piping and the heat exchange fluid thermosyphons to a pre-heating cylinder. It therefore works whenever the sunlight — or summer cloudlight — is warmer than cold water. The problem with solar heating is of architectural integration: avoiding nailed on appendages! My own house utilizes a seasonally-integrated system of solar power, back-boiler in a cooking-heating stove and small-scale hydro-electric power, any surplus not being used by appliances going into water or space heating. Few people are lucky enough to have their own electricity from wind or water, but terraced houses with party walls easily save the 1.3 kW that my generator provides. This sort of energy is something wholly physical and so

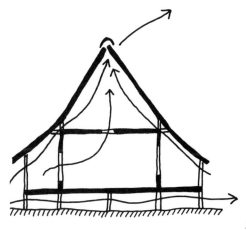

Cooling airflow induced by building shape.

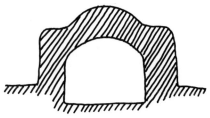

High thermal capacity to even out daytime and night-time, and even seasonal temperature variations.

Wet, heavy snow slides off.

Light, dry snow stays on the roof for insulation effect.

Although the need for space-heating is greatest when sunlight hours are shortest, the sun can make a significant contribution even where it must be supplemented by other forms of heating. This system (near Stuttgart) utilizes air circulation to heat radiant walls and floors (architect: Walter Bauer).

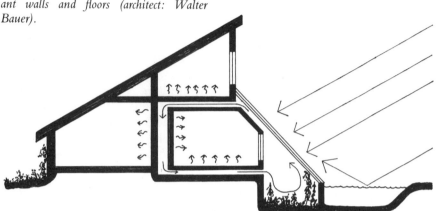

often considered only in the material sphere.

Concentration on sunlight as energy makes it easy to overlook its disinfecting and health-giving functions and its profound effect upon human moods. Just as many grievances between people result less from *what* was said than *how* it was said, *quality* can be more important than material description.

Tables of comfort/temperature can take no account of whether heating is radiant — in which case the air should be cooler — or con-

Building around a heat source: this stove chimney heats downstairs and upstairs rooms, reducing heating needs to a third of the average. Such planning is complicated by another requirement of minimum energy design: the need for compact space. Spacial economy requires central circulation spaces with rooms around them, whereas heat economy requires central chimneys!

vected. Forced air convection may create the right *temperature* but not *comfort*, for duct friction can reduce the ionized content of air by up to 95 per cent and adversely alter the balance between positive and negative ions. Synthetic materials producing high static electricity have similar effects — the serotonin content of blood is altered and other physiological effects are alleged, but the scientific evidence is conflicting.[3] Many people, however, experience headaches and lethargy when the ion content is low and positive ions predominate. The air is cleaned of dust and we feel healthier and more fully alive with a higher content and a negative to positive ratio of 60:40.

Electronic devices can make the air 'fresh as a mountain stream' but at the price of noise and electrical effects. Where outdoor pollution is low, opening a window normally suffices, but in polluted locations, in larger buildings where uncontrolled natural ventilation can cause draught problems, or wherever there may be special health needs (as for dust-borne allergy sufferers), flowing water can be used. The freshening effect of fountains in exhaust-polluted city squares can be enhanced by inducing particular qualities of movement in the water, and Flowforms have been developed for this purpose.

To be adequate, fluorescent light must be brighter than incandescent. It still saves energy, unless it is often switched on and off; if however you get out of the habit of switching light off, the fluorescent tubes themselves may be more economical, but lighting energy as a whole may go up! The physiological effect of normal fluorescents, with their sub-visible mechanical flicker, evening out light and shade, and restricted colour spectrum, can have health effects: headaches and eyestrain are the most common, though metabolic and social dis-

Flowforms are specially proportioned vessels designed to induce rhythmic oscillations in streaming water so that left and right-handed vortices combine in figure-of-eight movements. They attempt to enhance the constantly spiralling movements natural to flowing water, which help keep it fresh and invigorating. This picture shows a Flowform cascade incorporated in a handrail.

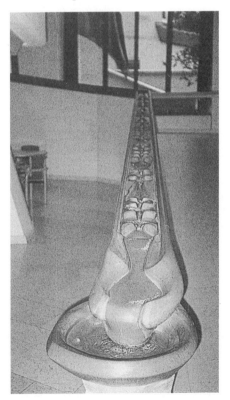

[3] Planverkets rapport 77, *Sunda och Sjuka Hus*, Stockholm, 1987, p. 111.

orders also result.[4]

Undue concentration on one-dimensional themes, such as warmth, light, acoustic absorbency, ease of cleaning and so on has tended to ignore what effects things have on the human body. In the last decade the concept of indoor pollution has rapidly advanced from obscurity to popular consciousness. We all know that inadequately ventilated buildings fill up with old breath, body odours and tobacco smoke. A Danish study of 15 offices *without* health complaints, however, showed this to account for only 12 per cent of total indoor pollution.[5] Many places are much worse.

What we now recognize as 'sick building syndrome' is not just a matter of inadequate ventilation. It has micro-biological, chemical, thermal and electro-biological dimensions. None the less it is something that anybody can recognize and simple commonsense measures can largely avoid. Sick building syndrome is taken seriously because it causes absenteeism due to sickness and has therefore economic implications. More serious illnesses can result but tend to receive less attention. It would be easy for instance on reading the Building Regulations to infer that formaldehyde is little worse than an irritant vapour released from urea-formaldehyde insulation. It is in fact a highly reactive poison and carcinogen and is present in some varnishes and many glues — and therefore glued products like chipboard, plywood and furniture.[6]

I do not wish to imply too materialistic a picture of simple and inevitable cause and effect. The human being is after all a being of spirit, not merely a responder to physical laws. A poisonous environment is no more certain to cause disease than a pathogenic bacteria. Not everyone who breathes, for example, *Bacillus legionellus* will become ill.[7] We develop illness, as I have mentioned, because to make inner transformations we need to experience a change of state. If our environment is more stressful than therapeutic it encourages this change-of-state to be found through illness. This unfortunately is often the case.

For the human body as well as the ecological systems of our planet there appears to be a general weakening of the immune defence systems. The increase of allergies is insidious but unspectacular: one hay fever study indicated a doubling within a decade.[8] Allergic reaction is now the greatest source of illness in Western society, affecting one third of the population,[9] and there is a growing speciality in medicine which looks for environmental

[4] Joachim Eble, 'The Living Language of Architecture', (lecture at conference), Järna, Sweden, 1986; *Environment Digest* No 9, February 1988; David Weatherall, *New Light on Light*, Management Services, September 1987, p.30.
[5] Birgitta Bruzelius, 'Dold olf bov i sjukthus', *Byggforskning* No 3, Stockholm, April 1988, p.17.
[6] *Ibid.*
[7] Only some 1-7 per cent will (Dr de Nutte, *'Indoor air quality'* paper at *Sick Buildings and Healthy Houses* seminar, London 22/23 April 1989.)
[8] *Sunda och Sjuka Hus*, p.197.
[9] Kerstin Fredholm, *Sjuk av Huset*, Brevskolan, Stockholm, 1988, p.51.

causes for allergenic conditions. The most extreme I have read about was an American doctor who wears a gas mask to interview his patients in case they are wearing perfumes he is allergic to! Many factors are involved in the growth of allergies, but research in Sweden where disproportionate growth appeared in the north with cleaner outdoor air but super-insulated houses has strongly implicated buildings,[10] even when pollen is the trigger agent, as pollutants have synergistic effects. When a police station had to be closed because the police were suffering from skin rashes, the view that buildings only affected the hysterically oversensitive needed substantial modification.[11]

All buildings modify human environment from what has been 'natural' over millennia to that which is a recent biological experience. It isn't only the characteristics of buildings as inert solids that affect their occupants — how they are designed, what they are built of, how they are built, maintained, furnished, heated, cleaned and ventilated — all are significant and involve owners, builders and users as well as manufacturers and architects. The health effects of materials and design are complicated. Modern building materials range over some 70,000 chemical combinations,[12]

releasing perhaps 1,000 chemicals to the indoor air.[13] It is known that there are synergistic effects, as for instance between pollutants, temperature, electricity and ionization of air, but there are also countless chemical combinations with as yet unknown consequences.

Another problem is that effects such as headaches, irritability, hyperkinesis, learning disability, fatigue, dermatitis, asthma, rhinitis, 'flu mimic conditions and irritations of the bronchia, mucous membranes, throat and eyes can initially be dismissed as nuisance irritations or be mistaken for 'normal' ill-health.[14] In fact they are but early warning signs of much more serious illness in the longer term.[15] Cancers tend to appear only many years after exposure to carcinogens, often more years than many products have been in existence. In short, we just do not know enough about what buildings do to us.

There is a desperate paucity of relevant and reliable information — particularly in English — in these newly emerging areas. There are specialist consultants, but they are too expensive for the sort of people, such as charities, that I work with. I often therefore have to resort to an intuitive assessment. Intuition, the zone beyond the frontier of personal knowledge, isn't reliable but

[10] Esko Sammaljärvi, 'How to build a healthy house', in Det Sunda Huset, p.119.
[11] Kerstin Fredholm (see note 8).
[12] Kerstin Hejdenberg and Inger Sävenstrand, 'Allergiutredningen i Sverige', (in Dawidowicz, Lindvall and Sundal (eds), Det Sunda Huset, Byggforskningradet, 1987, p.34.)

[13] Kerstin Fredholm. Sjuk av Huset. Brevskolan, Stockholm 1988 p.14; Jaakola and Heinonen, p.50; Sammaljärvi, pp.120,132, in Det Sunda Huset, p.89.
[14] Finn Levy, 'Sykdommer assosiert med bygninger', in Det Sunda Huset, p.89.
[15] Johann Högberg, 'Kroppens varningssignalar pa toxiska effekker av kemiska ämnen', in Sunda och Sjuka Hus, p.140.

can be refined to make it more so. We have a good basis for intuiting what is healthy or unhealthy in terms of whether we feel well or ill, or whether something smells, feels, or tastes good for us — a sensitivity we can develop. If you have ever sawn, chiselled or nailed arsenic-impregnated wood (e.g. 'tanalized' or 'celcured' timber) you will know how dead the wood feels, sounds and smells to work.

Nowadays, very few people spend less than three quarters of their lives in buildings or vehicles. Modern buildings are quite different in materials and climate from these of even only a generation ago. Health problems associated with dampness, cold, draughts, lack of sunlight and overcrowding have almost disappeared; the problems of toxins, radiation and electricity are to a large extent entirely new. New materials, new construction and new standards have brought new and hitherto unanticipated problems — so much so that when in 1971 the state laboratory for inspection of foodstuffs in Geneva moved into a new building suddenly all foods examined had unacceptably high levels of toxicity. Gases from paints, plastics, chipboard, etc. were poisoning the room air sufficiently to contaminate even the food samples.[16]

Up until the last war, buildings were normally constructed of 30-40 per cent organic, 60-70 per cent inorganic, but natural, materials (such as bricks and lime). Many nowadays are of 90-100 per cent artificial, synthetic materials.[17] Synthetic materials are often the cheapest, most convenient to use or have the best *material* performance — but they can be harmful to health. In fires many, especially plastics, become killers. In many modern building fires, smoke can kill in seconds.

It is not only vapour emissions that have biological effects: the way inert materials are put together do too. Rats, according to experiments, decline in fertility after three generations in a Faraday cage, likened by some to a reinforced concrete building.[18] When, in wartime Burma, the rumour got around that anti-malaria tablets caused impotence, troops refused to take them. Tower-blocks have got off more lightly! In multi-storey buildings, the biological effects due to terrestrial radiations filtered through underground (or piped) water currents, and those of the earth's geo-magnetic grid, are progressively amplified by each reinforced concrete storey until some consider it unsafe to live above the eighth floor.[19]

All materials shield to some extent against atmospheric and terrestrial radiations but some, notably plastics and metals, do so more than others. Sunlight and the moon's effect on tides, animal and human behaviour and plant development are all easy to observe, but what about all those other radiations that we can't feel or see? It's

[16] Vassella, 'Third Skin', *Permaculture Journal* 14, 1984, p.23.

[17] Michael Schimmelschmidt, unpublished paper.
[18] Vasella (see note 16).
[19] Joachim Eble, unpublished lecture, Järna, Sweden, 1981.

not just that after millennia of life in natural surroundings we are not adjusted to cosmically isolated environments; all life lives at the meeting of cosmos and matter, and more than a few miles from this meeting, nothing lives. Humans, unlike birds or earthworms, live at exactly that meeting point, with our feet on earth, our waking heads in the air. If we reduce these influences we are reducing the life-renewing, fertilizing power and health-giving balance of the marriage of earth and cosmos.

The meandering lines of medieval streets were not entirely random. An infra-red study of Regensburg[20] showed that the streets followed the lines of subterranean water courses thereby ensuring that the houses avoided them. Terrestrial radiation, ionized by passing through water running in friction, is thought to have harmful effects ranging from insomnia to rheumatism and cancer. Underground currents and waterways can be located by dowsing, which it would appear our ancestors undertook before building. There are other benefits of these curved street forms which I shall go on to discuss later but at the moment we are only considering the effects of architecture on biological health.

Many modern buildings have radiation levels some 25 times those of the external environment. Of documented nations, Sweden suffered amongst the most heavily from Chernobyl, yet on average people receive ten times more radi-oactivity from their own houses.

Radon gas is a decay product of uranium. This is present in materials of deep-earth origin and being incombustible is concentrated by burning so that materials like pumice, blast-furnace slag or pulverized fuel ash (of which most insulation blocks are made) contain significantly increased amounts. It degrades to products which can attach themselves to house dust, in due course being breathed and remaining some time in the lungs.

The average radiation content in a detached house of insulation block is 200 becquerels per cubic metre of indoor air, although (exceptionally) lightweight concrete *without* any ground radon contribution can give levels of 800 Bq/m³.[21] At this level death from lung cancer after 60 years in such a building is 80 times more likely than death in a fire, yet radioactivity is not mentioned in the Building Regulations whilst precautions against fire are given 88 pages.

Radon gas from the ground can disperse in the open air but is trapped within buildings. 'Action levels' (over 400 Bq of radon daughter products/m³ of air) are limited to only a few areas of Britain, but low levels are more widespread. Ground radon concentration can be reduced by vapour-sealing buildings from the ground (in so far as it is possible). Well-underfloor-ventilated suspended ground floors are better.

In addition to siting, choice of

[20] Joachim Eble, unpublished lecture, Järna, Sweden, 1981.

[21] Swedish figures, *Radon i bostäder, Socialstyrelsen, statens planverk, statens strålskyddinstitut*, Sweden, 1988, p.3.

materials and careful building to ground design, these effects can be minimized by reducing dust. This can be achieved by good natural ventilation and by avoiding fan-assisted or convected air heating, hot surfaces over 45° C (which activate convection),[22] dust- or fibre-producing or electrostatic materials, and dust-trapping design.

What are broadly classified as electrical diseases include cancers, miscarriages, metabolic misfunctions and allergies to electricity that are so extreme they can make normal life impossible.[23] We are now exposed to electro-magnetic radiation some 15,000 million times as strong as that reaching us from the sun and of every frequency from 50 Hz to 7,000,000,000 GHz. Only recently has concern focused on the health consequences of upsetting the electrical, electro-chemical and electro-magnetic balances of the body. Short-term symptoms include headaches, weakness, disturbed sleep, nausea and loss of potency.[24] Typically, more serious effects take years to show up, but various studies have suggested that up to 15 per cent of cancers in children[25] and 90 per cent of infant cot deaths[26] are related to electro-magnetic fields; VDUs have been found to increase the risk of miscarriage or foetal abnormalities by 53 — 80 per cent.[27]

Rubbing on synthetic carpets, wallpaper, paints, handrails and so on can charge people with static electricity, perhaps up to 15,000 volts.[28] If this is a negative charge it will attract dusts and mineral fibres to the skin causing rashes and eye irritations. In addition, static electricity speeds up the ageing of air[29] so contributing to a range of under-oxygenated blood ailments like depression and lethargy.

If synthetic clothing can produce an electrostatic charge sufficient to detonate explosives,[30] think how it is to paint a house with synthetic paint and fill it with synthetic furnishing, foams, carpets, bedclothes, veneers, and so on. In the modern office environment there are in addition VDUs, fluorescent lighting and multi-storey networks of complex underfloor cable.

The proliferation of electro-magnetic emittors may be something no individual can do much about (other than not buying microwave ovens, car-phones, etc.) but we can minimize the effects by locating buildings, in particular

22 Esko Sammaljärvi, in *Det Sunda Huset*, p.120.
23 *Dagens Nyheter*, Stockholm, 5 November 1987.
24 Dr Leslie Hawkins, 'Health Problems arising from geopathic stress and electro-stress', paper given at *Sick Buildings and Healthy Houses* seminar, London 22/23 April 1989.
25 *Environment Digest*, August 1987, p.6. The increase in computer games where children sit right in front of the screen can only make this worse.

26 Roger Coghill, 'Power Frequency Hazards', lecture at *Sick Buildings and Healthy Houses* seminar, London 22/23 April 1989.
27 *Environment Digest* 14, July 1988, p.6, and Dr Leslie Hawkins (see note 24).
28 Lasse Sjöblom, in *Byggforskning* 3, Stockholm, April 1988, p.28, Sammaljärvi (see note 22).
29 Sammaljärvi (see note 22).
30 British Army allegations that this causes premature detonation of IRA explosives must, however, be regarded with considerable suspicion.

housing, away from transformers and power lines (including railways) or the zones between them and large bodies of water. Within buildings spur cable layouts in metal conduit can be routed in less occupied areas at least 1m from water piping and from sleeping, sitting and working positions. Sensitive zones, such as bedrooms, can be switched off completely by 'demand switch' circuit breakers as far away as practical. Dimmers and night-storage heaters in bedrooms, not to mention electric blankets, bed-end television sets and the like should be avoided as the sleeping organism regenerates and is at its most sensitive.

The broadening of concern from occupational to indoor-environmental health consequences has been slow to develop. Public attention is now beginning to focus on wood preservatives, formaldehyde and pathogen-breeding air-conditioning systems, but to a large extent the conventional architectural world (in Britain at least) disregards any other problems. The military, however, take these things seriously: certain types of light and sound have health effects which have been developed as weapons, allegedly already in use.[31]

In many people's eyes a modern house is more healthy than a damp, mould-growing, draughty old one. Not that these problems can't occur in new construction: concrete takes three to five years to dry out, but buildings are often occupied immediately upon completion. The real difference is that health hazards in modern buildings are much less visible. In the old days, when buildings were draughty, of breathing fabric and made of earth and plant materials, even radiation from granitic rock had insignificant effect. Draughts are uncomfortable. They also add up in some cases to be the equivalent of leaving a door open; after minimal thermal insulation draught-proofing is the most cost-effective way to save money. Nowadays we therefore try to make draught-proof buildings. However, less draughts can mean less ventilation, causing in turn condensation both on surfaces and within the building fabric — the former can lead to mould growth, the spores of which are harmful to breathe, and the latter can threaten the life of the building, for instance by dry rot. Vapour barriers[32] solve this problem. Consequently, virtually every modern timber-frame building has an indoor environment wrapped in a plastic bag. Modern masonry techniques with foam-filled cavities, impervious paints or vinyl wallpapers (giving off poisonous vinyl chloride) are little better.

It's bad enough to wear plastic clothes, but to live in a plastic bag means that we breathe a lot of undesirable stuff: outbreathed air, tobacco smoke, gas cooker or heater fumes, dust and mineral fibres, and toxic vapours from building materials, fittings and furnishings. A lot of fresh air could solve these

---

[31] Kim Besly, 'Electronic warfare', *Peace News* 7, March 1986.

[32] Usually in fact 'vapour checks', because they are imperfect barriers. For brevity I use 'vapour barrier' to include both 'barrier' and 'check'.

problems but it means either foregoing comfort or excessive heating bills. The other way is to eliminate the plastic bag — to build buildings that *breathe*.

Buildings can be seen as the third human skin (skin is the first, clothing the second). The skin performs many functions: it breathes, absorbs, evaporates and regulates as well as enclosing and protecting. A building which through its fabric is in a constant state of moderated exchange between inside and outside feels — and is — a healthy place to be in. It has a quality of life. A sealed-fabric building is full of dead air.

Some rocks have infinitesimal vapour permeability. Build them into a house with strong cement mortar and you have, effectively, concrete. Lime-based mortar on the other hand has a greater permeability and capacity to absorb vapour. If you demolish old granite and lime-mortar farmhouse walls, the moisture content of the core of the wall can be seen to differ little from winter to summer. Such walls have a very high moisture moderating capacity.

Lime was once shellfish[33]; timber once trees; bricks once clay — that mineral, with its curious colloidal properties, nearest to life. Plastic and steel are long removed from any history of life, and buildings of these materials have no moderating, breathing, living effect on the internal environment. Timber on the other hand, as well as

moderating humidity by virtue of its extensive internal air spaces, absorbs dust particles and airborne toxins.[34] If not sealed (by paint or polyurethane for instance) or poisoned with wood preservatives, it is one of the healthiest materials to live within. Conventional timber construction, however, requires vapour barriers to protect the building fabric, even though this penalizes life within by allowing indoor pollution levels to build up to many times that of outdoor levels.

To enhance the life of indoor air I try to find safe ways to dispense with vapour barriers. While some of the many buildings which should have them, but don't, show as yet no ill-effects, mistakes can threaten the whole building, or by causing mould and fungal growth can make for an indoor environment less healthy than a sealed one. None the less there are techniques. Reversal of current wood-frame construction (where timber is externally clad with brickwork) by using a slow-permeable masonry wall *inside* insulation and rear-ventilated timber or tile cladding can obviate the need for a vapour barrier. Such techniques must follow the rule that the vapour permeability of materials (or construction) must increase from inside to outside. Materials which, without deterioration, can absorb, hold and diffuse water are an advantage. After all, thatch has provided waterproof insulation without the need for vapour barriers for

---

[33] Much of it certainly was; there is some dispute as to whether all of it was.

[34] Vassella (see note 16). Hartwin Busch, 'Biologically safe building fabrics and the "indoor climate"', lecture at *Sick Buildings and Healthy Houses* seminar, London, 1989.

thousands of years.

Adequate ventilation, even on roofs where ceilings are on the underside of rafters and without vapour checks, can disperse condensation vapour sufficiently.[35] A membrane may however be desirable to obstruct mineral fibres or vapours from their phenolic bonders or from foamed plastic insulation materials from entering the room space, and indeed it becomes necessary whenever the balance between moisture generated, moisture-tolerance of materials and adequacy of ventilation is suspect. 'Semi-breathing' construction, where the roof or wall itself is vapour-check protected but an out-ventilated cavity is formed on the room-side with vapour-permeable material, can overcome this problem.

Cobolt Arkitekter in Norway have developed a system whereby buildings can be designed to breathe in more than out, so that as little interstitial condensation as possible occurs. A 'stack effect' ventilation path, achieved for instance by having an open fireplace, can draw air in through the walls. If it is of sufficient vapour permeability and short cuts have been eliminated by draught-proofing, moisture-rich internal air cannot enter the wall because outside air is coming through it. Such a wall saves energy in fact because heat leaving the building is warming up the incoming air. Even using protective wind-shields, vapour permeability must be limited to take account of gales,

and so will at times be inadequate to prevent airflow from inside to outside. Particular attention therefore needs to be paid to the moisture-moderating properties of the building fabric as well as to the absence of any pollution source within it (such as formaldehyde from chipboard or plywood). Because it has to take account of ever-changing conditions of weather and occupancy this sort of construction is complicated to gauge. Conventional vapour-sealed construction on the other hand is simple to gauge — simple because it is uni-dimensional, lifeless.

This issue of life-full or life-less is at the heart of architecture. Much construction is designed to protect the building fabric itself, not the inhabitants or the living surroundings. When a building is suffering from woodworm or rot, is it better to poison the occupants along with the infestation or use less guaranteed methods that are biologically safer? Organic solvent preservatives give off vapour in the short term; working on buildings sprayed a few months ago, you can notice the effects, and over many years future occupants are exposed to skin contact and breathing impregnated dust. Alternatives do in fact exist, but they are less easy — woodworm can be heat-treated at 50-60°C, but I only know of one

[35] The building regulations, HMSO, 1985, artF2, p.697.

[36] Synthetic pyrethrum-based Permethrin has low mammalian toxicity to protect bats, whereas Lindane remains toxic for 30 years. Borax-based sap-diffusion (e.g. 'Timbor') impregnated timber is imported from Finland but I have been unable to obtain it. Biofriendly preservatives are, at long last, beginning to appear on the British market.

contractor in Britain who offers the service. Less toxic chemicals such as those based on borax or pyrethrum can also be used.[36] The Building Research Establishment regards destruction of dry rot (serpula lacrymans) by poison as merely a secondary treatment — more essential is the elimination of conditions and existing fungus.[37] The poison however comes with a 30-year guarantee: more responsible methods depend on the architect's insurance!

If I use ordinary preservative-treated timber I now have to recognize that if the offcuts are burnt I will produce dioxin-laden smoke or leave arsenic in the soil. If I use any industrial product I will be party to pollution from its manufacture, some of it horrendous — I think of the shiploads of chemicals that are incinerated in the North Sea — some less so, but still bad enough

[37] 'Dry Rot: its recognition and control', *Building Research Establishment Digest* 299, July 1985.

In many city streets, the pedestrian zone is the most polluted. It is here that we need pollution absorption and redress by plants and active water. These we need also for health of spirit.

— I think of landscape eaten away by quarries. In the days when environmental concern first became fashionable I saw an advertisement for wood-grained plastic laminate entitled 'Save a Tree'. But how many trees have died as a result of air-borne pollution exported by the chemical factories that make such products? Our responsibilities must, like the effect of our buildings, extend beyond the boundaries of the site.

Once we think about *ecological* consequences we must consider a whole world of relationships such as the effects of even a single building on micro-climate, flora and fauna. In the old days, barns even used to be built with entrances for barn owls. This sort of symbiotic consideration has almost disappeared. When we cover an area with buildings and paving what does this do to air quality? In cities you can notice the air differences between parks and paved areas. To replace the oxygen breathed by one person needs 1½ sq m of grass or a 5 m diameter tree crown.[38] If we think of buildings as supporting vegetation and use approachable, locally appropriate materials, in general we reduce distant pollution effects. In the built environment, as in the landscape, the more harmoniously balanced the ecology, the more artistic the effect.

The built environment is shaped by the decisions of many 'responsible' professionals: planners, architects, surveyors, engineers and many others, including business executives. Yet the results don't show it! In general our built surroundings make people feel less well, less at peace within themselves and less able to cope with biographical crises rather than resolving them as inner turning points.

Another way of looking at it is to ask to what gods are these decision-makers responsible. In the choice between building-life or heating costs against occupants' health, economic factors are weighed against life in the same way that governments balance health-service expenditure against taxation. For all their faults, governments at least *have* to think about this choice. Most of us, however, never bring this issue to consciousness. But when we don't think about things other forces take over.

Whatever our role, architect, builder, building owner or user, we are all amongst the shapers of the human environment. Our work will not support life, unless we consciously choose that as a priority — whatever the price.

[38] Andreas Engelhard: 'Solararchitectur im Industrie- und Verwaltungsbau' in *Gesundes Bauen und Wohnen* 1/90 March Nr.38 p.38.

Flowform in a pedestrianized city centre.

5

Qualities and Quantities

One definition of architecture is 'the design of buildings'. One definition of buildings is 'durable enclosures of controlled environment' — the creation of environment appropriate for certain (usually human) functions.

Out of this approach has grown up specialized areas of study. Environmental science is concerned, on the whole, with quantitative descriptions of what is appropriate for the physical needs of human activities, such as the temperature we need to feel comfortable in when we sit at our work or exert ourselves physically, and how much light we need to read or to work in a kitchen. These quantities tell us nothing of what is a *nice* atmosphere to read or cook in, but this nice atmosphere is made up of the right warmth and light and so on. It is this atmosphere that gives meaning to the quantitative physical descriptions. To create nice and, more importantly, meaningful, appropriate atmospheres we need to focus our attention not on the quantities but on the qualities.

There are instruments to measure quantities — for instance how loud something is and of what frequencies the sound is made up. We have tables with which to evaluate these quantities: noise over so many decibels disturbs sleep or intrudes upon conversation, and so on. We use instruments to get objective information, but they are selective. A wind rose from anenometer readings can tell us where best to site a windmill, though it needs editing to remove over-speed gales. For siting buildings, however, we need to know the temperatures and humidities of the constituent winds, which ones really matter and which don't. Several instruments, a computer and the patience to wait several years can tell us this, or we can ask some of the older neighbours. Instruments unquestionably bring an objectivity beyond that we can discipline ourselves to, but their selectivity can mean that their answers often do not correspond to our questions. What is so often dismissed as human subjectivity is the unconscious ability to synthesize many factors; however, because it is un-conscious many personal prefer-ences get muddled in.

How can we find a similarly objective basis to evaluate qualities? It is possible to quantify human responses on the basis of the aver-age of answers to questionnaires or by recording how laboratory rats behave. It may be that there are meaningful ways of doing this, but I myself have been put off on the one hand by the over-simplicity and obviousness of conclusions which we all know anyway, and on the other by the fact that I am trying to be human and not a rat-like responder to behaviourist stimuli. When we are in places that cause us to feel good (or bad), it is likely that our feelings are shared by other people. I do not mean buildings or places which I *think* something about, but those where before I start to think I *feel* something.

When we look at the evidence of sick buildings it becomes clear that qualities which are widely liked or disliked, be they heat, light, sound or whatever, are in fact beneficial or harmful: our subjective preference judgement can be very meaningful as a guide to whether places are good or bad for us. For some scien-tific measurements in fact the human being is the best instrument. In developing the Olf and Decipol system of measuring indoor air pol-lution, Professor Fanger found human assessment to be more sen-sitive, accurate and meaningful than any chemical analysis.[1]

The problem is that there are *per-sonal, cultural* and *universal* layers of response. Normally, these are all muddled up together — we just respond; we don't think why. It is however possible with disciplined exercises in dispassionate objectivity to start to distinguish them. If we can see past the personal, we can start to use our *own selves* as *objec-tive instruments* to evaluate qualities of environment.

The amount of space between ourselves and strangers that we need to feel at ease in varies from culture to culture.[2] Quantitative space requirements are predomin-antly cultural — to Europeans for example the scale of American houses, cars and cities is striking. We also have personal spatial prefer-ences: some like the cosy, some the grand. In absolute terms, however, distance affects how we need to speak, move or focus our eyes. It therefore has bodily effects, regard-less of our expectations and prefer-ences, which influence social relationships.

Colour is highly personal. We each of us have colours which we prefer to wear or cannot stand. But there are also fashion colour cur-rents which flow through society around rocks of established conven-tion. There are also *universal* aspects of colour: red speeds the metabo-lism, blue slows it down. This is a physiological fact — everyone responds this way. Different colours stimulate different glands: for in-stance, yellow — thyroid, blue — pituitary, red — male sexual, vio-let — female sexual glands.[3] Knowledge of this kind can be used

[1] *Byggforskning* No 3, April 1988, pp.17-19.

[2] See for instance Edward Hall, *The Hidden Dimension*, Doubleday Anchor, 1966.

to manipulate people and can also be used therapeutically. In a home for maladjusted children in England there is a swimming pool illuminated underwater so that the children's splashing bodies can appear coloured: red helps activate autistic children and bring them out of themselves into activity, blue helps calm down the hyperactive ones and bring them into themselves.[3]

All colours have universal effects. I don't mean little blobs of colour, but whole *experiences* of it — coloured light, coloured environments, walls and ceilings of colour. Heavy, strong colours have a tendency to be too forceful to be comfortable with, and their use requires great skill and sensitivity. Traditionally they are applied in a variety of hues and shades utilizing harmony and counterpoint. Strong colours have a tendency to be manipulative — they dominate the furniture and other oddments and also the human being. They *force* their mood upon a room.

Where colour works as a delicate breath, however, is in the light. Coloured light has a different effect from pigment — with light you can feel raised up into a mood, but with pigment pressed down into it. Imagine (if you can't arrange to experience it) the room you are in bathed in yellow light or painted yellow — or blue or red.[4]

Except for special rooms for special uses, coloured glass tends to look out of place, especially as most windows are for looking out of. We have to find other ways of influencing the light-mood of rooms. So called 'lazure' is the technique of painting thin transparent veils of pigment on a textured white ground so that light reflects through the colour, each veil of which is so thin as to be barely visible. Translucent veils of curtains can have a similar enlivening effect on the colour of light.

Green is a colour of balance; it has a peaceful, calming, soothing effect. (In Steiner schools, it is the balance colour for classrooms at the mid-point of childhood.) Yet it requires considerable skill to paint a room in opaque green without it becoming heavy and dead, for green is such a lifeless colour to paint with. Worse than that, there is the risk that reflected light will green people's faces creating, by association, a disquieting mood. If on the other hand light shines in through foliage it can be both life-filled and peace-bringing.

I use light reflected off natural materials a lot. For this I generally depend upon white walls and ceilings. Where a specific colour mood is appropriate I use lazure. These colours are not random, not just heart-warmers or attractive, but are specific for their function. What colour for instance would aid the transformation of a cafeteria from a utility-food canteen to a centre of sociability? What would be the colour to prepare us for entry to a church? This is not a matter of rules, but of *cultivating awareness* of how

[3] Kenneth Bayes, *The Therapeutic Effect of Environment on Emotionally Disturbed and Mentally Subnormal Children*, Gresham Press, 1970, p.31.

[4] Experiments on physiological and psychological reaction to colour are, in fact, more often carried out with light than pigment. Kenneth Bayes (see note 3), p.30.

colours speak. The next step is to bring the colour into conversation with the light — unique to every room — and then to work with the ingredients artistically.

To work with the qualitative vocabulary of architecture we need to cultivate this awareness in all spheres. We need to experience more consciously — not just think about from outside — that which it is all too easy just to float along through. We need to wake up our senses, the gateway between reality and our feelings. The senses tell us about what is important in our surroundings; mostly, we experience things through the outer senses: sight, smell, taste, sound, warmth, touch. Architecture in the sense of environmental design is the art of nourishing these senses.

We try to make places that *look* good. Even if we don't look *at* them, the background visual impression is one which causes a good response. People see this visual 'mood' — they can talk about it afterwards, remember it for years, but when you ask them to draw any of it they have hardly any idea how it actually looked! Much of this visual mood is made up of colour, visual texture, scale and the quality of meetings between things. Much of our response is due to the quality of light.

It doesn't matter how nice a place looks if it smells of bad drains. The smell of fresh bread or ground coffee can be a shop's best advertisement — better than any visual display. Salesmen sometimes look oddly at me when I ask, will your product (say carpet) smell? They look even more oddly at me when I sniff their samples! It's no good designing a place that looks nice but smells horrible, especially as that smell means something about the air we breathe.

Adults don't go around biting their surroundings — but babies do. When one sinks its teeth into a plastic or wooden toy it gets quite different tastes. When we taste copper or lead in drinking water, for example, we start to wonder if the pipes are poisoning us.

Warmth can have such different qualities: radiant heat from the blacksmith's forge can be bearable even in the summer, but even in cold winter warm air heating is unpleasant. The focal radiant warmth of a stove or fireplace, reinforced by the sound, smell and sight of the fire, gives a spirit to a home. We call this part of a building the hearth — a heart. Anyone who enjoys a hot-water bottle or lying on sun-warmed rocks can imagine the luxury of Russian stoves which are built to sleep on. What a difference between conducted heat and air-conditioning!

Most of us do not go around deliberately touching buildings, yet without thinking about it we touch them all the time. Textures which we walk on or feel with our hands (or eyes) make all the difference between places which are approachable and which are not: not many people would prefer a concrete bench or steel table to a wooden one. Few of my clients ask for any particular material in any parts of their buildings, yet a lot of them ask for wooden floors — often, unfortunately, where I can't arrange adequate underfloor ventilation.[5]

How rooms sound — whether they echo, resonate or absorb — can make all the difference to their mood. A church, a living room or a restaurant should *sound* different, and materials and physical design can be arranged to achieve these effects. We don't feel at home in hard echoing rooms. A dead acoustic space is not good to sing in. Noisy clatter turns a restaurant into a canteen.

These are the outer senses. They are our contact with outer reality, what in the East is called *maya* — illusion — although through them we can see beyond this into the invisible spiritual reality that lies behind it. We also have finer senses with which we can perceive this invisible reality — and it is *very* real. We experience it for instance if we go into one shop where the object is to make a lot of money or another where the main object is to provide a socially beneficial service.

We can cultivate our sense of what a place *says*. We can begin to sense the unspoken values that lie behind the outer phenomena that are manifest in the way it has been planned, the way it has been built, the way it has evolved, is cared for and used.

We can cultivate our sense of the individuality of places — not just the outer differences but the differences of spirit between places. Much of this is manifest at a lower level in the extent to which different sensory experiences reinforce or contradict each other. But we are

playing with cosmetics when we design with these surface phenomena alone. Places really speak through their spirit of place, and the phenomena accessible to the outer senses are consistent with that spirit. Mass housing, system designed, system built, imposed upon the landscape, isn't going to feel a great deal better if it is painted attractive colours, or if the road noise is screened. It still remains environment for statistics, not for individuals.

We also have senses which tell us about our own state: our states of physical balance and of movement. In particular we have the least conscious sense of all — a sense of health. It is exceptional for you to *feel* healthy, but you do feel it when you are *not* healthy, when you feel ill, thirsty, craving a cup of coffee, tired and so on. This sense has a lot to do with architecture because a lot of buildings can make you feel ill, even in the short term. In addition to causing identifiable symptoms, they may affect our general state inducing, for instance, bad sleep, tension or exhaustion.

On the whole, however, the outer senses can give us good guidance as to whether an environment is harmful or health-giving to the physical body and the human spirit. Unfortunately the senses are so undermined these days that unless we cultivate them, they may speak too weakly for us to understand their message.

For proven physiological reasons, people can feel ill if they work all day in artificial light. Yet the light of spring can bring such joy to the heart, it can get the invalid out of

[5] There are, I know, approved timber floor constructions without underfloor ventilation. None the less I do not entirely trust them.

bed! Inadequate light can cause Seasonal Affective Disorder, associated with depression, lethargy and suicides. Yet too much light in a room requires it to be too open, unprotected — and we do after all build buildings for social and environmental protection.

Several smaller windows are better than one large one, not only because, from the energy-saving point of view, for the same heat loss there is a better distribution of light, avoiding quantitative extremes, but also for quality. The light is more full of health-giving — and aesthetically satisfying — life. Also you get two views instead of one, which helps you to orientate yourself: I have been in enough buildings where I didn't know which way round I was! Even for occupations which theoretically have no need of natural light, windows offer a contact between the artificially-controlled indoor world and the weather and life-renewing cycles of nature outside. Improvements to view-out in hospitals have been found to reduce post-operative recovery times and the need for pain-killing drugs.[6]

Nasty smells warn us that something is bad for us. Smell is borne by minute quantities of matter, but even minute quantities can be harmful. The lungs have a huge surface area — we breathe several thousand gallons of air per day. Even if you don't believe in homoeopathy or the Bach Flower Remedies (or for that matter silicon chips, depending as they do on minute

impurities), you can see that a small amount of poison in the air can have very significant effects. All materials that smell are giving off vapour — plastics for instance, for all their longevity as rubbish, are not absolutely stable; they give off vapours of plasticizers, stabilizers, pigments and unattached monomers.[7] It is these we smell.

Generally the effects are subtle. We have to cultivate the ability of our senses to tell us what is good or bad for us: when we touch polyurethane-coated wood we know there is something wrong. It feels hard, smooth, cold; it does not breathe and the finger's sweat condenses on its unyielding surface. It *looks* like wood but it is a lie and it is hardly the best food for the human spirit to surround it with lies. If you want to bring children up to be honest it is not going to help if their environment is full of lies. Nor in the barrenness of its sensory experience does it nourish the soul.

When we talk about nourishing the soul, we are talking about finding qualities in the environment that provide the right balance to the imbalance of the moment. Of course, there are a lot of imbalances and a lot of soul needs, and a few are major ones. Sometimes we lack society, stimulation, sometimes we have too much and it is stressful. Sometimes we need to withdraw to a secure private domain such as a fireside, inner garden, or personal retreat. Can we find fulfilment of

[6] David Wyon, 'Buildings fit for people to live and work in' in *Det Sunda Huset*, p.196.

[7] Katalyse Umweltgruppe und Gruppe für ökologische Bau- und Umwelt Planung: *Das ökologische Heimwerkerbuch*. Rowohlt (Hamburg, 1985), p.197.

such needs in our surroundings?

The harder and more lifeless our surroundings are, the more tired, tense and sapped of life we tend to become. The softer and more alive they are the more renewed, relaxed and healed we tend to be. Soft lively air rather than rough funnelled draughts,[8] absorbed sounds rather than hard echo, moderated enlivened light dancing perhaps off water or through leaves from different windows with their ever-changing interplay of subtly different light and shadow. Vegetation brings softness, life and seasonal rhythm. Indoor plants not only soften architectural hardness but (ferns especially) can redress the ion balance in the air. Plants outdoors can be used to moderate micro-climate. They give oxygen and life to the air we both pollute and breathe. Central Europeans wonder how British industrial cities survive without trees, but traditionally the British rely upon island sea breezes, generally from the west, to blow pollution across the eastern parts of cities and the underprivileged who live there. Other countries incorporate wooded 'lungs' into urban regional planning.[9]

Vital as they are for air quality we enjoy trees and other plants for their restful appearance, life-filled shade, leaf sounds and scents. They are breath for the soul as well as the lungs. Climbing plants can not only soften hard corners, make unyielding textures approachable, enrich walls and clamber or cascade in archways, but also absorb street noise. There can, of course, be problems with vandalism, security, wind-damage and maintenance, but often the problem is to find the will and energy to get plants going. Some vegetation requires no maintenance, but few clients take me seriously when I suggest spraying buildings with cow manure to encourage moss and lichen.

There are people who dislike trees. To them they are just slippery leaves on the pavement or robbers of window light. These are sound objections (however much we might like to dispute them) but they are usually out of proportion to the benefits vegetation can bring. When we meet such objections, however, we need to be able to consider honestly whether trees in these particular locations would be a benefit. I certainly know places where they would not, though they are few.

Even vegetation can be the wrong thing in the wrong place. It's much easier to buy cultivated than wild varieties. Often these have been bred for one-sided development like coloured leaves or extended flower season. These may look all right in the garden centre but quite out of place when planted.

We can even have too much vegetation. Much as I like them for aesthetic, landscape-harmony and ecological reasons and have been asked for them on a number of occasions, only twice have I had the right projects for turf roofs.

We can also have too much light, too much heat — we certainly know when we don't have enough. There are lists of what temperature

[8] Gusts caused by high buildings kill some 200 people each year.
[9] See for instance Design with Nature, Ian McHarg, Doubleday/Natural History Press, 1971.

As well as improving air quality by oxygenation, humidity regulation, ion breeding, dust absorption and smell masking, plants can soften the impact of hard materials and shapes, harsh light and acoustic environment. Their transformative effect is far out of proportion to their cost.

and how much light we need for various activities. Of course once qualitative things are described solely in quantitative terms, they must be a little bit suspect: degrees Celsius regardless of whether the heat transfer is radiant, convection or body contact, lux regardless of whether sunlight, incandescent or fluorescent lights — quite apart from any aesthetic aspects. But on the whole we can be guided by these standard lists: they describe what is appropriate to whatever physical situation.

We can have too much of something or too little — after all, all life on earth lives in a very narrow band between earth and cosmos, between absolute matter and the heat of the sun. Healthy life is always a delicate balance between extremes. Architecture supporting this health is also a narrow band, containing a whole world of qualities appropriate to different states of being. If we step outside this we become manipulators.

All activities have the need of particular soul moods. Our environment can have qualities appropriate to what we are doing such as views to give space and peace when our work is tense and claustrophobic. Practically, these cannot always be distant views over calm water; sometimes they must be of wind-stirred treetops or of changing clouds. Or perhaps we need warmth, enclosure and focus, as when we sit around a fireplace — not the place for windows unless they have interior wooden shutters to close out the outside.

We need qualities appropriate to our mood of soul. At different times, we may need exciting, socially stimulating places or relaxing, calming places to be in, places perhaps bathed in rose, greenish or bluish light, not heavily painted but light-bathed as by light filtered through leaves, flowers or coloured glass or reflected from walls lightly painted with transparent veils of colour.

When we are exposed to stressful situations, it is as great a support to experience peaceful surroundings as it is to find a place of cosiness and warmth on a bitter winter's day. At other times we need to be a part of human vitality: young people in particular need this as part of the process of stretching their social horizons. Most specially we need root-giving experiences in today's rootless times: the stable institutions of former generations — marriage, employment, social order — can no longer be taken for granted. All of us live under a lurking threat of ecological or economic collapse, even nuclear extinction. If ever we needed stability, peace and roots in our environment, we do today.

We also need qualities appropriate to our present state of development. Is it perhaps a bit paternalistic to talk of what *other* people need? However, if we don't offer these qualities through conscious choice the world will be dominated by currents that are not chosen for the benefits and freedoms that they bring, currents from outside the individual like the manipulative pres-

While roofs shelter building occupants, they rarely have a beneficial effect on local outdoor climate. Vegetated roofs can however absorb airborne toxins, redress oxygen, ion and humidity balances and also reduce rainwater overload of storm drains.

sures of advertising or behaviourist psychology, currents from within like egotistical desires.

Choosing what is *appropriate* is not determining or manipulating but offering an environment supportive to balanced development. Inner freedom depends upon this balance. What, for example, is appropriate — and nurtures the development of inner freedom — in an environment for teenagers? This is a delicate question because it involves many factors: teenagers need to throw off parental moulding influences and peer-group conformity provides support during this period. On the other hand this conformity is fruitful ground for economic exploitation, especially as it is bound up with their newly-found experience of their bodies, sexual desires and emotions of unfamiliar power. It is easy to hide the tenderness of this new emotional self behind acting tough and insensitive. Conformity can easily pass from being a support to an obstruction to that life of commitment to ideals which marks the entry into adulthood.

Different qualities of room, be it schoolroom, café or youth centre, will support different aspects of this growth process and nurture different pictures of the developing human being. Depending on our picture, or at the very least transitory image, of the human being, we would design quite different rooms. One room would perhaps be dominated by harsh centrally-directed lighting; the harsher the better — in some cases stroboscopic, multicoloured or on the edge of ultraviolet. Around this and in the shade would be many protected niches with seats and small tables. There would be a drinks counter, invitingly lit. This room would have the character of being underground — if possible it would be — there would be no windows and entry would be by a narrow passage. Spatially, it encloses, but with non-assertive surfaces. Preferably the room should be non-rectangular, its ceiling lowish, and painted in strong, dark, opaque colours — brown, black, sometimes dark red. You are not aware of all this because the lighting focuses attention on people moving. It is a room of the night with the privacy of darkness contrasted to dramatically-lit self-exhibition. With loud music this makes a successful discotheque, also a certain sort of café.

A dance hall for people conforming more with their parents' than their peer-group convention is quite the opposite: electric candelabra, high ceilings, rectangular and more apparent spaces, broad access-ways. When the morning sunlight streams in through the open fire-exit of the discotheque, the place seems hollow, disenchanted, fake. It is only an environment for *part* of the human being. When the sun streams in the ballroom, it doesn't make so much difference. This is an atmosphere which seeks to tone down the different aspects of the human being.

If on the other hand we try to design a different sort of room, if in teenagers we wish to encourage questioning social awareness, we need more natural light: windows with views of *things going on*, perhaps sitting or sprawling window-alcoves which are not private from the main space. The architecture would have cleaner lines but not simplistic forms. It may have gentle curves which are more open, less protective. We would seek light-enhancing pale colours, perhaps blue-grey veils enlivened with tints of green. At night we could expect areas to be gently differentiated in light, heat and comfort: a light above each circular table set in a corner and surrounded by seating, cushions on the floor near the fire, hand-made music. Such a room would foster a society based on friendship, on the whole human being.

Similarly, a room for seventeen-year-olds would be quite different from one for five-year-olds. They are not only taller, with higher eye-level and more interested in the world beyond their immediate surroundings, but also more active intellectually. Five-year-olds need a more protective environment, opportunity to live actively in a world of imagination and imitation. One room would be more upright, firmer in its forms and spaces, more outward-looking. The other would be warmer in colour, softer, lower, snugger in space and form. For the kindergarten I am building at present I have circular rooms with their singular social focus but freed from their deterministic geometry by corner play-alcoves at various levels and large windows as well as tiny low deep-set ones with protection offered by a world of trees.

Different environments are appropriate to different social groups. It is not just a matter of cultural responses — what is spacious to someone from Bombay is intolerably overcrowded here; what is wild landscape in the Netherlands would seem like a metropolis in Lapland. There are also the classic differences between what sort of home is sought by people who work in urban environments, alienated and stressed by crowds of strangers or by outdoor-job people such as farmers, foresters and seamen and others who are battered by the weather. One group needs space, peace, light, air, long views and tend to indulge themselves in private realms — suburbia; the other needs cosiness, enclosure, protection, and have tended to build solid houses with small windows, sleeping in tiny cabins, cupboards or high-sided drawer-beds.

I know old Welsh farms where the farmers have become prosperous and built new extensions for spacious kitchens — for the kitchen is the room everyone lives in — but in the modern, sterile style. The house seems empty, unfriendly: the family complain that it feels cold, though of course it now has central heating. The lack of warmth is not because the temperature is too low. It is not because the colour of the walls is too cool a white, though this is part of it. It is because the house has lost its heart. It has lost its soul warmth and now has *inappropriate* qualities.

This failure to nourish the soul is experienced also as a failure to provide the right physical environment, even though the instruments say otherwise. The qualities of environment are more important than their quantity.

6

Conversation or Conflict?

We can concern ourselves with the appropriateness of sensory qualities to the needs of a place and its users, but very few qualities can be considered in isolation. Most, although they work upon us anyway, we *consciously* notice only by contrast. We notice the warmth when we come into a warm room from the cold, when we move closer to the fire. We notice the smell of a city or industry when we first arrive there, the next day we don't. The wrong sort of warmth or air quality is harmful whether we notice it or not, but if qualities, however appropriate, are to bring joy and refreshment to the soul, we need variety — not endlessly the exact correct temperature, lighting level, the same view, the same sort of shapes, space, or movement through space. Once there is variety we become aware of how one experience is set against another. We become aware of *meetings*. Mostly it is in the visual sphere that we notice meetings. Most of these are meeting edges, for while the *being* of something may live in its centre — say a field of

colour — the meetings occur where it meets another colour. This edge between them can be hard, or so subtle you can only say where the centre of each colour is; the rest is just 'somewhere in between'. *How* they meet makes all the difference!

In every aspect of life there are two extreme ways of meeting: conversation or confrontation. In one you are open to what the other brings, in the other you seek to impose your own pre-formed viewpoint. Whether we are talking about relationships between individuals, groups within society, nations or power blocks, confrontation leads to polarized positions and seeks to resolve matters with force. Regardless of who wins or loses it is a destructive process. The loser is oppressed, the winner demeaned.

Conversation is the process by which two or more individuals come together to create a whole *more than the sum of their parts*. They must listen to each other and to what comes into form through — and only through — the conversation. The individuals need to be able to adapt their plans according to the needs of each other but *without compromising their essential nature.*

Listening is much harder than it sounds, because one needs to put one's own thoughts aside for a while. Adapting without compromising is easier than it sounds, because uncompromisable principles, if *honestly* and *morally* founded, will not be incompatible with each other. The conversational ideal can be a light to guide one in human relationships and daily life. It is the essential foundation for harmony. Neither socially nor in any art can you build living harmonious relationships out of rules. They depend upon listening responsiveness. If architecture is to provide harmonious surroundings in which people can feel alive, at ease and peaceful enough to feel themselves, it needs to build out the conversational principle.

A lot of time in architectural design is given to creating *shapes.*

Elevations are drawings of shapes, many of which don't exist of course because they are unrepresentative planar views. There are also many other shapes which do exist as drawn but which we don't see because we are unable to stand back and see everything all at once. What we *do* see a lot of is *edges*, outlines, corners, openings through solid walls, meetings between planes.

Shape, whether we consciously look at it or not, has effects upon us, but *how* that shape is edged also has a great effect. Television screens are basically rectangular. All except

Built in the 1950s this room was a severe and sterile rectanguloid. Minor shaping of the ceiling, hand-finished texture on the walls and the interplay of light from different windows make all the difference. This minimal shaping makes the room habitable. The mood however, is created by the light, the colour and the reflected colour-light. Space, shape, light and colour all weave in conversation with each other to create one atmospheric whole.

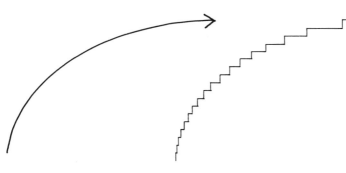

With microdot display the steps may be almost concealed. When however I want to alter my hand curve, it evolves into a new   *living whole, whereas the computer curve just adds and subtracts bits.*

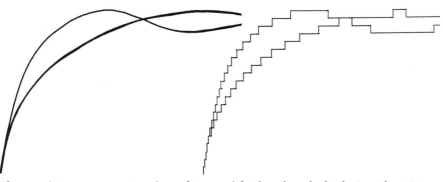

If we try this we can experience how of two apparently identical shapes, one is filled with   *life, the other absolutely is without it.*

a very few modern ones have the corners rounded off, and I doubt that people could watch them for the many hours that they do were it not for this.

Subtle modifications to shape to ease the movement of the eye (and hand) from one line to another make a tremendous difference to how we respond to things. A table with knife-sharp edges is not as nice to sit at as one where the edges are rounded — but the effects can be more than just niceness.

Imagine a small white room, almost square, one high-level window only, no view — a monk's cell. Softly undulating plaster, a subtle curve on the ceiling and above the window; the clay-tiled floor laid on not quite straight lines; the sunlight enlivened on the uneven surfaces of wall and floor.

Imagine it again, the edges knife-edged, the walls shiny smooth; ceiling, walls, floor meeting each other in hard, precise lines; the sunlight a sharp rectangle.

The first is a room for prayer, a place of tranquillity set aside from the hubbub of the world. The second somewhere you cannot wait to get out of. Anyway, that's how I feel about it.

For practical reasons, especially construction and storage, we need the straight line and its product — rectangular forms. But these are not forms which we can find anywhere in the human body, in human

However striking, sculptural or imposing, architecture like this is the product of the rational but arid intellect, not the heart, for it seeks powerful images at the price of more delicate feelings. Such buildings do not create places to feel good in, nor indeed, with hand, eye or heart, to feel at all. Nothing can live in a hard, rectangular, mineral world without artificial support — without vehicles, lifts, air-conditioning, TV, muzak, consumerist entertainment it would be uninhabitable. It is no coincidence that New York City uses as much electrical energy as the whole continent of Africa. Places resulting from this approach to architecture are for machines, not for the human soul. What sort of a picture of the human being lies behind this approach and what sort of person does it create?

movement, human activity, nor anywhere in nature. Rectangular forms are forms that suit machines and mechanistic thinking. An excavator has difficulty digging a curved trench; a quantity surveyor finds it hard to estimate a three-dimensionally curved surface.

If I use a computer to draw a curve that I have swept with a light-pencil it can of course do it — but the nature of the two curves could not be more different (see drawing p.59). If I draw a curve, it has bodily movement built into it; the most alive and firmest curves are drawn from the toes, not just with the fingertips. To display this, the computer reduces my uninterrupted, flowing, evolving, living gesture into a lifeless binary code.

Things that are alive never fit exactly in any hard-edged category, as I am reminded each time I fill in a computer questionnaire. Design by using coordinates on a rectangular axis may make things easy to measure, but it describes the world in the same unequivocal way that the computer's yes–no code does. It is no accident that a world made up only of rectangles is death to the soul. Hard mineral matter, hard lines, hard corners, repetitive unambiguous form. Nothing can live in such places if it were not artificially sustained to an immense degree. Along with other things this abstraction and artificiality feeds alienation so that you can walk with your open eyes blank past an accident, past a cry for help.

The straight line is — we know — the shortest route between two points. In other words it has only *one* concern — not the most gentle, most lively, but the *shortest* route. Machines make *dead straight* lines. The feeling hand — if it is not trying to imitate machine standards — makes *nearly* straight lines. They are as different as the movements of the clock and the movements of the universe: one dead, one alive.

When you shave the edge of a piece of wood with a plane you get an edge which results from the geometry of the tool. The straightness is imposed upon the material. When on the other hand you shave it with a knife or chisel, the tool responds to the grain and knots. (One needs to be very subtle with this or things can look a bit 'ye olde'.) Rooms made up of sharp planed edges are harsh, those of responsive knife-shaved edges much more life-filled and welcoming.

When the Bauhaus movement so enshrined the geometric solids cube and cylinder, it was also choosing the most economical forms for the machine age. Bauhaus derivative buildings that followed were invariably concerned with monetary criteria and the forms extruded by machines were highly suitable. Structurally the straight line is a line of tension, tension between points. It therefore can give firmness to forms, but the ends of the line are crucial points. What happens at the end of these lines? Usually one line meets another at right angles. The point is now doubly strong — but it is worse than that! We orientate our life in three great planes: forward, vertical and horizontal, the product of these axes — in front-behind, above-below, side-to-side. These axes have completely different

Conflicting lines, planes and shapes can be brought into conversation, even song, with each other. Try not to see the picture but to imagine the experience of going up these stairs, turning, and passing into the room beyond. Now try to imagine it in an environment of right-angled meetings of unsoftened corners and dead-straight lines.

characteristics: one is the axis of time (past and future), one of surroundings, and one of oneself, standing in tension between cosmos and earth.

Horizontal and vertical could not be more opposite in structural principle. Dead-front to side could not be more opposite in terms of human movement in or around a building. When these planes meet at right angles, their different characteristics at the maximum, their meeting is without give-and-take or metamorphosis. It is forceful but dead. While acute angles are uncomfortably compressing and obtuse angles invitingly embracing, the right angle has a stable balance. Where we can bring life to this balance we can benefit from its organizing characteristics, but if we cannot its mech-

Traditionally the corners of stone buildings have a two-angle flare. In the first 70cm (2'6") of height it reduces perhaps 15-20cm (6"-8") then again another 15cm (6") reduction in the next 2-3m (6'-10'). It is easy to describe this but less so to do it. It needs feeling hands to create the quality of a rock bursting through the ground rather than a dropping lump of jelly.

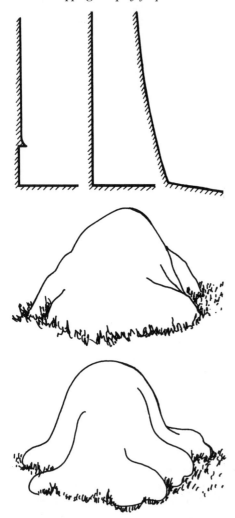

anical, lifeless, qualities will dominate. Even furnished diagonally across the corners, as they often are, rectangular rooms with rectangular doors and windows and hard smooth surfaces are not places for *living* occupants, only for mechanical bodies. Staying in such rooms I experience them as uncomfortable, claustrophobic and life-suppressing.

If the meeting between planes can be mediated then their different characteristics can be brought into a poetic relationship, just as words are as suited to poetry as to the parade ground. Textural softening or breaking-up of the lines and planes helps. In conversions of old buildings I am often stuck with vertical walls, horizontal ceilings and rectangular rooms. Undulating plaster and exposed beams overhead transform such rooms from boxes to attractive places. It helps even more if I can break away from the exactly vertical. A wall is more solid, rooted in the earth, more timeless, quiet, reassuring and restful if its base is wider than its top.

Classical architecture always grew from bases, vernacular walls invariably widened at the foot, usually with a two-angle flare. Only in modern times have we become used to vertical walls rising straight out of the ground without base or interceding element. I don't use a uniform taper but increase or swell it at the base to obtain firm, strong forms, varying the angles and gesture to suit the circumstance. I also like to have a little of this quality internally. These buildings *belong* on the earth whereas others which meet it ver-

How buildings meet the ground makes all the difference to whether they belong in this particular place or have merely been parked there.

tically are only *parked* here.

Of all meetings, how a building meets the ground is perhaps the most important and yet the one most commonly unconsidered. A vertical, right-angled meeting (or worse, a concrete pillar) takes no account of any rooting meeting between ground and building. One or both needs to be shaped to the other.

The claustrophobically-entrapping quality of a harsh meeting between ceiling and wall can be transformed into one of welcoming enclosure if the ceiling can rise a little. Shape matters even more in a building's 'eyes', those frames through which our view or we ourselves pass in and out — windows and doors. Even the slightest curve in the top of a window makes the frame softer; it is also a structural shape — in brickwork an arch, in timber where the beam overhead passes from cantilever to span.

I use a similar approach to meetings of planes in plan — where I am not able to avoid rectangular meetings I can place, say, corner furniture to intercede between two walls. Our ancestors did it all the time — a corner cupboard in plan, skirting boards and mouldings in section. Often this softening of a harsh meeting needs only very subtle treatment. Triangular corner blocks in the tops of rectangular windows only 1x3cm or a plastered edge finished by hand instead of a straight-edge can make all the difference. But even with the most subtle modification of shape a gesture starts to be implied. If the corner block at the window head is horizontal it is quite different from if it is vertical. Does the ceiling arch only gently or reach down to the wall?

Three straight lines together can

There is a lot of difference between a straight line which suddenly swings into a curve and one in which the curve is implied in the straight and the firmness of the straight in the curve. If your eye journeys with the lines, the quality of one is of unrelated steps; the other weaves both past and future qualities and has a dynamic life in the straight, an organizing firmness in the curved.

How different this street and corridor would be if they were straight.

imply a curve — even more so if the corners are rounded and those lines are soft, perhaps even gently curved. The straight and the curved have markedly different effects. One gives firmness, orientation, the other life, fluidity.

An example of life-filled, life-enhancing curves is the movement of water in a mountain stream. Here all the lines and movements have a *breathing* rhythm. This has a rational, physical basis made up of the interaction of forces — gravity, momentum, friction and so on. It is in fact a manifestation of the interweaving relationships between the different elements of water, air and earth. The ever-changing pictures we see in the sky are made up of a different combination: air, water and warmth.

There are also curves where the formative forces have no such enlivening conversation: a squiggly line, although it has come out of the human fingers, has no other inner formative principles. All these curves have a certain dreamlike characteristic: we can sit and watch a stream for hours; the squiggly line is remote from any practicality — a good line to doodle with.

Free curves, in other words, are life-enhancing but they can lull us into a dreamy state, forgetting the practicalities of the real world and our tasks to work for the social good. In contrast, straight lines are lines of organization, often of power, but they are life-sapping. A balanced, healthy human life lies somewhere between these two extremes.

Sometimes we need more firmness from our environment, some-

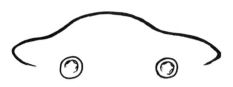

The environment provided for us is dominated by straight lines, whereas objects that compete for our attention frequently utilize curves to enhance consumer appeal.

phous forms, making forms for instance which are visibly bound by the principles of gravity or accelerating-decelerating curves which have a strength and vigour which arcs of a circle do not.

The balance between organizing principles and fluid life-forms itself needs to be not just one quality measured against another, but at every moment a weaving into a single whole. A patchwork of dead and amorphous pieces, like alternate functionalist and organic buildings along a street or a jungle indoor garden in a geometrically severe atrium, is no conversation.

times more fluidity — but never one entirely without the other. Not compartments all straight or all curved, but both sewn together. We need therefore to give life to the firm geometric — especially the rectilinear elements — and to give firmness to the non-straight — especially the fluid forms. The former we can do by moderating the *meetings* and the *planes*, the latter by bringing structure-giving principles into otherwise amor-

We live in a world swamped with rectangular buildings. There are also a lot of man-made curvilinear forms around us: consumerist artefacts (most notably stylish cars) use curved lines for their appeal. Our environment — especially if we live in cities — is made up from sterile rectanguloids and deceptively persuasive, manipulative, curves. The provided, dependency-making and predominantly practical world is characteristically rectangular. The

Both curves and straight lines are one-sided. They need each other, need to be able to weave their positive aspects together. Not just added to each other (A) but rather, the straight (firmness and organization) in the curved and the curved (life enhancement) in the straight (B and C).

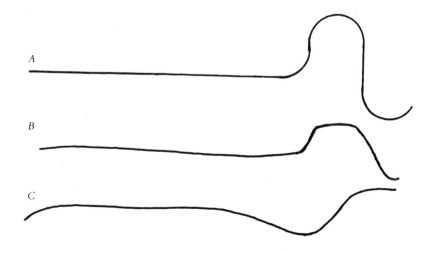

Part of the charm of traditional places is that even though made up of rectangular elements, *the lines are not dead straight (houses in Lavenham, Suffolk and Kano, Nigeria).*

As a general rule, it is easier to make a firm curve out of straight lines and a life-filled straight line out of curves (that is, drawn by the hand and arm which tend to radial movements). For example:

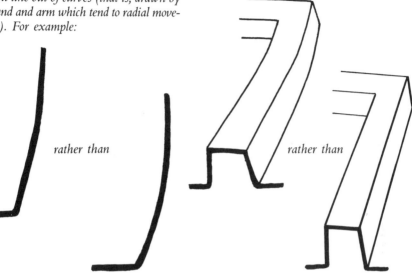

rather than   *rather than*

Different planes and angles can be brought into relationships where they converse *together as one single whole.*

The simplest conversation between windows and ceiling shape is when they reflect each other's shape. We can go on to develop one as a metamorphosis of the other so that a single organizing principle gives a different form in every new situation or material, much as the forms of landscape, cloudscape and current patterns on water can be metamorphoses of each other. Different elements can start to sing together as in multi-part harmony.

objects in it, competing in customer appeal and often chosen to express individual personality, tend to be fluid-formed. The schism of practical and feeling into polarity-developing reaction undermines the wholeness of the human being. Like all polarization this makes moderating harmony harder and harder; as with all reactions we become less and less free, for our actions result not from our own conscious choosing but from the thing we are reacting against. To support wholeness, freedom and health we need to find ways in which the essence of the straight and the curved can be meaningfully integrated.

Conversation is the art of raising disparate elements into one whole,

raising it above the levels each was previously trapped in. This applies to every sort of architectural relationship, the metamorphic experience between inside and outside, between one space and another or between adjacent visible elements. It also applies to the relationship between user, architect and builder.

To meet, elements often need to be modified in their form to respond to each other. A square window can look out of place against a sloping ceiling. As you pass from an arched corridor into a room with another shape of ceiling a metamorphosis needs to occur to bring them into a meaningful relationship — a metamorphosis of space and form that reflects the change of mood experience that you experience as you move from one space to another.

This principle can be brought into every form of meeting so that elements do not just collide with each other but *speak* to each other — indeed, so that they sing together.

Harmony in our surroundings is no mere luxury. Our surroundings are the framework which subtly confine, organize and colour our daily lives. Harmonious surroundings provide a support for outer social and inner personal harmony. Harmony can be achieved by rules — but rules lack life. Or it can arise as an inevitable but life-filled consequence of listening conversation. Even between the same group of people, different times and places trigger different conversations — even more so when the people are different. This one principle can give rise to many forms — not just the way I do it! This inspiration is so much at the centre of my approach to architecture, that I could not imagine working without it.

7

Space for Living In

We have looked at *how* things meet, but what about the things themselves and the space between them? What about the effects of different shapes of forms and spaces?

If you make forms in clay, pour plaster over them and scrape out the clay, you get *spaces* exactly the same as the original *forms*. It is possible, therefore, to define space as 'negative form'. But the spaces don't have the same effect as the forms had. In fact the experience of space is quite different from the experience of form. So what is form? What is space? What is the essence of their different characteristics?

If, in an empty landscape, there is an object — especially a man-made object — our eyes are drawn to it. Even something we would choose not to look at will draw the eye. Statistically, a factory chimney may be an infinitesimal part of our field of view, but its presence colours a vast landscape. In the same landscape, wherever there is a hint of enclosure, especially if this is a meeting point of different elements — for instance earth and water or

woodland and clearing — this is an inviting place to sit and eat a sandwich.

'Form' is the property of objects. Objects are, in a way, beings. They may be dead in themselves but their presence radiates influence all around them, some more so than others, like a formica table in an otherwise homely room. 'Space' is the space in which things can happen, in which living things can *be*. Buildings are objects from the outside, space within. Their presence is made even stronger by what goes on within them. Sometimes their form, uniformity or lack of fenestration conceals what goes on inside. Big buildings like this are like lifeless giants.

All over the world the era of well-intentioned urban renewal created threatening, lifeless monsters standing in expansive seas of no-space. There has been a well-deserved reaction against this approach from both architects and the public at large. Nowadays buildings tend to be more varied in their forms, but there is still a tendency to think that interesting individual buildings improve the environment. Architects think of buildings, look at buildings, react to buildings. For other people, buildings are *part* of the landscape; parts of buildings — the façades — are part of the townscape.

Things which happen inside and near buildings interest us; the buildings themselves merely set the mood, and this is so even for great architecture. People didn't go to the medieval cathedrals to look at the architecture but to take part in religious sacraments. It is similar for the buildings of the Renaissance or for modern times.

Yet to go to the opera in an unadorned prefabricated shed — however good the acoustics — would, to most people, be an inadequate experience. Architecture, in other words, sets the *mood*. It also provides space or boundaries to outdoor space *in which things happen*. It influences both the physical mechanics and the mood — the soul relationships. The mechanics of how we come into contact with people influence the quality of our relationships. However attractive the surroundings, cross-flow in a busy railway station, for instance, will aggravate stresses already caused by travelling. In a restaurant project I once had, my client wanted the washing-up sink *facing* the customers; the washer-up became part of the community, not an excluded menial servant.

Shape affects relationships. A circular table tends to bring people who hardly know each other into group discussion whereas a rectangular one with more than six seats tends to lead to separate conversations. A circle has a single focus. You can build a brick dome by tying your left hand — the hand that lays the bricks — to a peg in the centre. When you sit with your back to the wall in a circular room, you are as bound by the centre as was the piece of string with which the room was marked out. This is a good shape for community, for meditation, a shape to choose for a village community dance, an egalitarian discussion, a meditation chapel — but it is inflexible.

The Sioux Indians lived in circular family tepees arranged in a tribal circle, in a world which was conceived as a circle — a unity of spiritual forces. But the focus of the tepee — the fireplace — was not dead centre, nor is the plan of a tepee exactly circular. The circle of the world was intersected by a cross — the directions of the active beings of north, south, east and west. The rigid geometry of the circle was set alive. When government soldiers rehoused the Indians, first in square forts, then in rectangular houses on reservations, they severed them from a spiritual relationship to the world around them, manifest in their built social forms, and destroyed the roots of their culture.

If you walk around a square you experience abrupt and regular changes in direction. Orientated on polar axes this is a good shape to reorientate yourself if you have lost your inner bearings. It is firm, balanced, but differentiated by the light. This is the shape for curative eurythmy rooms.[1] Square rooms need to be set into life, for instance by shaped ceilings. Flat ceilings turn

The circle is the shape for meditative or social forms. To enliven the static geometry in this circular chapel, a subtle axis was introduced by this window grouping. Fortuitous circumstances presented this rock in the floor. We considered removing it but would the violence of dynamite be the right spiritual foundation for a sacred use?

them into boxes. In all cases, if they are unrelieved and too exact, squares risk the deadness of symmetry and repeated measurement.

From the purely functional point

[1] Eurythmy seeks to translate that which underlies music and speech into bodily movement. This can be strengthened in a space which makes visible the planes to which we unconsciously relate our experience of the world. Eurythmy can be developed in experiential or therapeutic directions or for artistic performance. Depending on the emphasis, the basic requirement of a balanced, organizing, measured, upright space would therefore be modified in one or other direction.

of view, right-angular forms have little reference to the forms, movements, mood-of-enclosure requirements and mobile thinking of the human being. I do not wish to condemn the rectangle but to try to work towards an experience of its being. There are far too many rectangles in our environment these days, *far* too many. Few have been consciously chosen as appropriate shapes, forms and spaces; most just to simplify design and adapt construction to industrially-processed materials and prefabricated systems.

Simple geometry emphasizes the effects of proportion. In a rectangular room the proportions are obvious to everybody; not so in a cave. There is nothing inaccessibly mystical about proportion. The classical architectural proportions are found in the human body, in nature, in the physical wavelengths

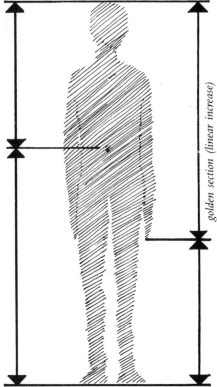

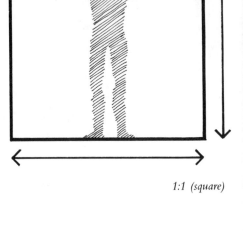

1:1 (square)

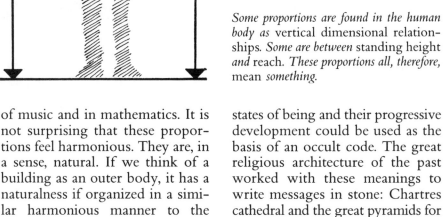

golden section (linear increase)

Some proportions are found in the human body as vertical dimensional relationships. *Some are between* standing height *and* reach. *These proportions all, therefore,* mean *something.*

of music and in mathematics. It is not surprising that these proportions feel harmonious. They are, in a sense, natural. If we think of a building as an outer body, it has a naturalness if organized in a similar harmonious manner to the human body. But what *is* proportion?

Proportions used at right angles to each other relate to different states of balance. Balance is something dynamically alive: it brings movement to rest, rather than freezing it. Proportions used in sequence relate to different kinds of growth. Both therefore in different ways are concerned with different states of being.

Relationships between these states of being and their progressive development could be used as the basis of an occult code. The great religious architecture of the past worked with these meanings to write messages in stone: Chartres cathedral and the great pyramids for instance have been so interpreted.[2] These messages were occult, for initiates only. But they were more than just the most durable kind of book imaginable; even today they convey special spiritual qualities, although perhaps we cannot say exactly what or why.

Even without understanding their bodily origin, cosmic foundation or spiritual significance, differ-

[2] Bodvar Schjelderup, *Evidence*, Forlaget Fredag, Trondheim, 1986.

ent proportions have a different feel to them. If one makes a building out of rectilinear elements, organized geometrically and freed from minimum space and functional criteria, it is not difficult to work with proportion. I myself do not work in this way. I therefore have to try to cultivate a *sense* of proportion, a sense I can use even when elements cannot be defined dimensionally, for instance because there are curves or sloping lines or because the eye sometimes reads the size of a window opening and sometimes the size of the glass only: it depends on the light.

Some things are definitely too low — cosiness can become oppression; some are too high — a room may be unsettlingly vertical; some are too broad — a view may be too open, making the room too outward-oriented for its function. Everything has its right measure of size and proportion. If I shorten the handle of a spoon by 5mm, broaden a window by 5cm, or raise eaves by 15cm, the qualities change noticeably. If I concentrate on the soul moods I wish to create in a place the proportional relationships suggest themselves. They are definitely right (or wrong) for the spiritual function: normally it is only afterwards that I can check them mathematically.

Sacred architecture as the measure of God's kingdom on earth is an architecture of proportion. Its degraded form, given a different impetus by mechanical-geometric thinking and industrial component manufacture, is the rectangular architecture of today. While the legacy of cosmic architecture could

have taken a variety of forms (as in Islamic designs) the main direction has been the establishment of a convention of rectangular buildings. In view of the materialist shift of philosophy and everyday thought from the Romans on, it is no accident.

A rectangular grid can order the chaotic. With vertical dividers to shelves we can bring calm order to previously confused clutter. The divine order of ancient architecture required such an uncompromising geometry to give it earthly expression. The Romans, increasingly concerned with order in the material world, used rectilinear grid planning for this order-giving purpose. Government-organized colonization of North America likewise used rectangular grids to apportion administrative districts, private landholdings and, in due course, roadways. Although arbitrary interaction of plan pattern with topography can be dramatic, such layouts have tended to foster similarly box-styled buildings set as regular blocks. Life-filled, stimulating, varied and harmonious forms are lacking. It is no coincidence that the USA (especially in the Sixties, the heyday of the rectangle) led the world in exotically curvilinear automobiles and hallucinatory drugs.

Throughout the developed countries of the world, dominated by material possessions, the dominant house form is rectangular. Indeed throughout history it is countries with rectangle-based built environment that have led technological development — the applied science of the materially practical. Rectangular spaces may not be life enhancing but they are, after all, the best shapes to store objects in. Organic filing cabinets are not so very practical.

These days most rooms are rectangular with hard smooth finishes. They don't even have the distraction of ornament or patterned surfaces, nor fireplace breasts, shelf-recesses, diagonally-set doorways, bay windows and elaborate mouldings with which the Victorians used to moderate their boxiness. Some rooms are indeed designed as boxes for storing people: you can read this thought in the designer's mind as you look at the plan or step into the room.

Such rooms are made occupiable

Sometimes building plans convey the impression that the architect's concern was the storage of people.

by their furniture, house plants, decorative objects and so on which draw our attention. You need quite a lot of things to achieve this. Minimally furnished, such rooms are not calm, holy, sacred places but empty, sterile boxes. Some rooms, by their function, may have shelves on all their walls. These, like shop windows, give interest through the variety of things organized on them. Other rooms need a lot of nice — and usually expensive — things to make their barrenness welcoming. Many rooms and gardens need ornaments with no practical purpose other than to make the place habitable.

It is not just that rectangular rooms are convenient to put things in; they *need* a lot of *things* to make them rooms we can live in. They are both product of and *fuel for* a

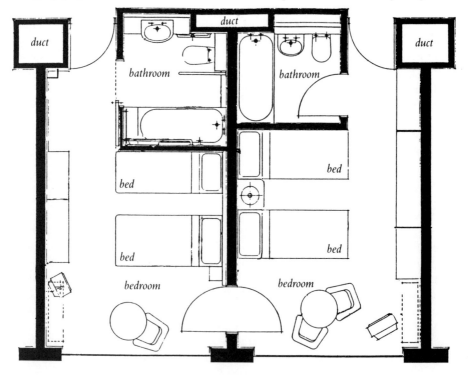

From the outside also, buildings designed this way are, in many people's eyes, stacks of people-storage boxes.

materialist culture.

I must stress that it is not the *rectangle* which is the problem, but *its life-sapping characteristics.* Where materials, textures, colours, light, living line and human activity can reinvest such forms and spaces with life, the materially practical and culturally-normal characteristics of rectangles can be used to advantage. None the less, in general, I feel on much safer ground with non-rectangular, or shape-moderated, spaces.

For all their sterile and materialist associations, rectangular buildings have been the mainstay of European architecture for a thousand years. They used, however, to be imprecise rectangles in plan with roof shapes which greatly softened the form. In modern times the rectangles, their surfaces and their rooflines have become harder and more geometrically pure. Most importantly, the spaces that rectangular buildings bounded were rarely exactly rectangular — even when they were they tended to be enlivened by sloping ground and subtle variations in buildings. A typical town square was more likely to be squar*ish* than square. The spaces, in other words, *the places where human life takes place*, were alive, not rigidly bound by a dead geometry.

Although geometrical principles underlie the forms found in living nature, straight, live forms do not exist there. Forms made up of straight lines meeting at right angles could not be more opposite to the forms of the natural world; they are, for better or worse, man-made forms. Their use in ancient sacred architecture is a language of man's relationship not to outer surrounding nature, but to the cosmic world, the world of geometrical principles.

In due course such forms became forms of status; a rich man's house or a craftsman's planed surface was set apart from a peasant's by precision and geometry. In our time, straight-line forms and computer-calculable geometry are the forms best suited to machine production. Without the elements of cosmic principle and careful craftsmanship these are forms which lack life. Wholly organic forms which nature surrounds us with on the other hand are forms which lack any imprint of human consciousness. Life-filled forms *for human environment* must lie between these two extremes, as does the human state.

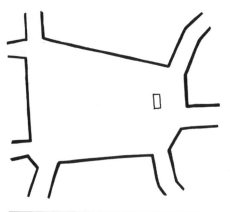

Before the era of consciously-designed town planning, a typical town 'square' was not normally geometrically square.

Places we live in and use are at least a step away from nature. Indeed, most buildings, even old ones with earth floors, are entered by a deliberate step. One of the difficulties of designing for wheelchair access is the loss of changes of level like this that make so much differ-

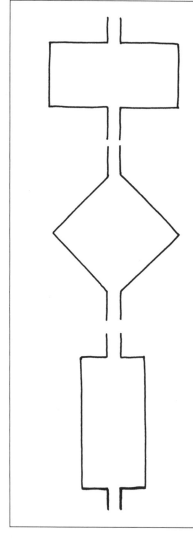

How we enter a rectangular space has a great bearing on our impression of it.

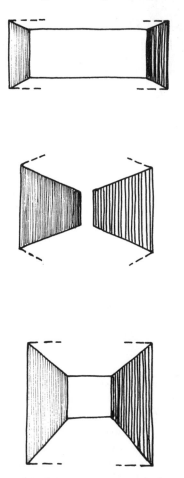

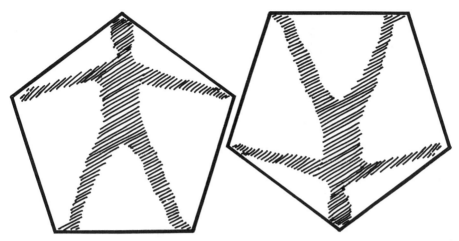

The pentagram: holy space or devil's form?

ence to how one place is distin-guished in mood and use from another. A large living room with kitchen and bedrooms opening off it can be a traffic cross-roads or a place of peace. It is changes of level, deliberate steps *up* to light and ele-vation and *down* to enclosure that make all the difference. Where the floor level cannot be stepped, I try to achieve the same effect with differences in ceiling heights, light-ing level, relationship to outside ground level and so on.

How we come into even a rectan-gular space has a great bearing on the impression we will have of it. Even more so when the space is non-rectangular. The pentagram is known variously as a holy space and the devil's form, depending upon which way up, or round, it is. I could never understand why until I tried to live into the experience of how we meet the shape. When we enter in the middle of a side we see an inviting enclosure. When we enter from one corner, it is con-fronting, symmetrical, crushing

individual freedom. Subtle perhaps? But if I wanted to make a facist architectural experience, say for a courtroom, I would enhance the pentagram as shown below. Try to imagine it. The experience is so powerful, so *un*free, that I wonder that it has not been used.

I am only just now beginning to experience the qualities of geomet-ric forms and spaces. For that rea-son I hardly use them, for they are dangerous to play with. I know that the great pyramids were forms of

When we enter a pentagram from the mid-dle of a side, the enclosure is inviting (top); from one corner, however, the rear wall con-fronts us.

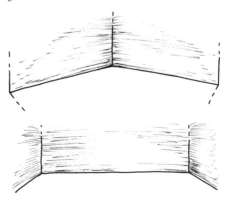

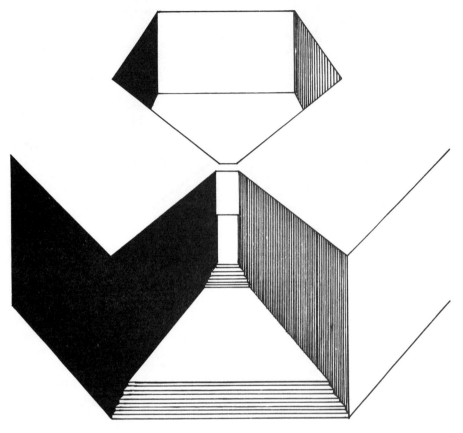

Above *The devil's pentagram can be developed into a powerful fascist space.*

Below *A very uncomfortable plan shape to be in can be made pleasant if the acute angles are blocked off to become obtuse. Now the room is a private enclosure, especially if we add a window asymmetrically as shown.*

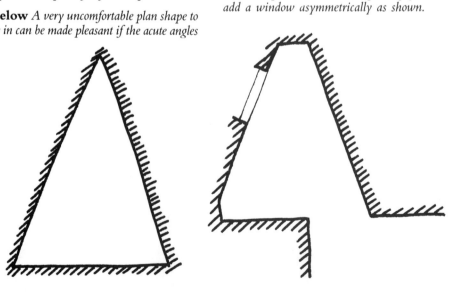

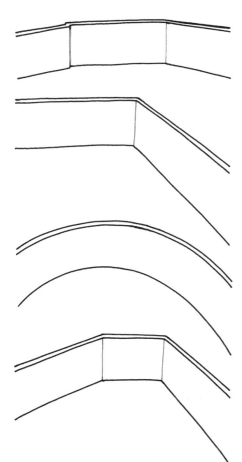

embalmment, that at the focus not only were stray mice embalmed, but the initiate Pharoah could experience physical death *without* dying as part of his initiation experience. I also know that people build these forms to sharpen razor blades or to 'recharge' them-

Angles wider *than the right angle are more welcoming. They tend however not to be very firm; as they bring little emphasis, they are difficult to bring into balance. If all the lines are the same length things are a bit dull. If one is too long, the shape easily goes out of balance.*

selves — but this is strong stuff to play with if you don't know what you are doing! I feel on much safer ground when I make up architecture out of simple walls, floors and roofs.

We have looked at the right angle.

To give firmness, such spaces often need one or two right angles as 'anchors'. This is not a rule, just something I often do without thinking about it.

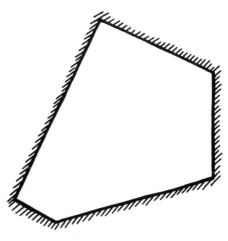

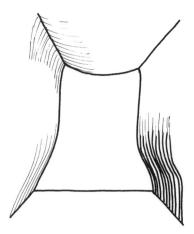

This is not a secure space. The ceiling weighs heavily — it sags. The walls droop. Not so bad in canvas perhaps — but terrible in concrete.

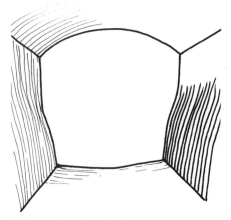

This, on the other hand, we experience as stable, enclosing, secure.

I now want to consider the quality of the *corner*. A corner can be at any angle, any curve. It can be welcoming and enclosing or excluding, uncomfortable and confining.

These are spaces in plan. We experience these shapes quite differently in section. What we need here are walls which bear on the earth, ceilings which enclose.

In different materials, such sec-

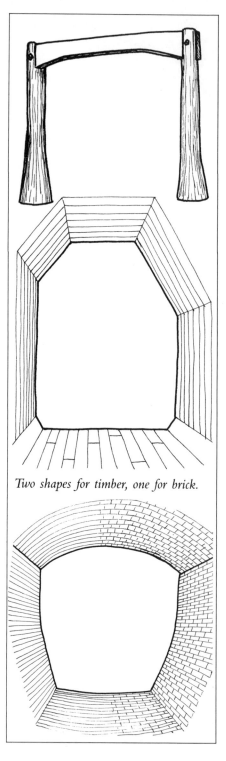

Two shapes for timber, one for brick.

tions would of course be formed differently.

All these shapes *imply* enclosing curves, yet only some are made up of curves. Others are made of straight lines, though preferably not 'dead-straight'.

We think in terms of lines but the eye does not see them. It sees edges where one colour or tone meets another. Drawing by making lines around things is no more than a code for what we see. Outlines blind our perception by replacing it with frozen concepts, yet architecture depends upon them. Design needs drawings and drawings need lines. Only the sort of sculptural forms that must be modelled and scaled up can do without lines. Lines make architecture — but what sort of lines? Some are complete abstractions produced by the edge of a straight instrument, some are lines that enliven the boundary they describe. If an architect draws a line it gets built whether it was drawn

with feeling or disinterest.

It is often so that even in the design of organically formed buildings it is almost impossible to avoid at least some more or less straight lines. Sometimes even for firmness I choose them instead of curves. But I want life and gentleness in these lines. When I use colour, texture, shape modification, interceding elements, corner cutting, elusive-formed surfaces and so on, what I am trying to do is to bring to the straight the quality of gentleness, harmony and life found in curves.

Whereas straight lines bring relationships of tension between clearly-defined hard objects, curves sing. They bring life and freedom — the mountain stream is full of life, markedly cleansing itself from biological pollution in a few miles. The water which flows in straight lines out of reservoirs, in canals or drinking water pipes has a reduced ability to sustain life. In some water, such as outlet channels from reser-

These lines emphasize the line and its straightness, the meeting is incidental. I was taught that architects should draw like this.

These lines feel the turn; the lines are only supportive to the qualitative key-point. As you draw it you can feel the shape and how it turns with your whole body.

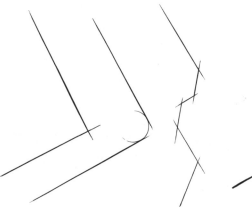

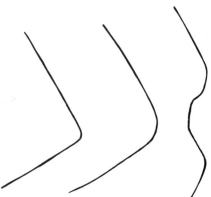

voirs or straight smooth-bore pipes, for reasons of shape of flow nothing lives.

This life-enhancing power of three-dimensionally curvilinear, rhythmically oscillating movement so typical of natural water flow, has been enhanced by the development of Flowforms, as described in Chapter 4.

Traditionally in China locations with harmonious combinations of qualities were identified as the wellspring of a health-giving force. This force was led across the landscape by systems of curvilinear lines both natural, like rivers, and man-made, like streets. Straight lines impeded this flow of health causing places to be unhealthy to live in. These were the sites to be sold cheaply to the Western colonialists.

But if all the shapes around us are too soft, it is hard to remember our tasks in the world, our tasks to do good deeds, to fulfil others' needs. This is the shape for the travelling people's 'benders' — round sagging tents of willow twig and canvas. Wonderfully enclosing, womb-like places to *be* in, but not places to want to work or concentrate or *do* anything in. For inner reasons we need firmness (but not to be dominated by it) — enough strong lines or their strength in strong curves, organizing geometry or structural principles to orientate our lives, to give daily life a framework.

What do I mean by 'strong' curves? If you draw an arc of a circle, it has a constant quality. If you

Sewage water has its life-bearing characteristics enhanced by flowing in rhythmically oscillating, three dimensional curves. It progresses through a series of 'flowforms' and open, vegetated ponds. What, conventionally, must be regarded as excessively one-sided biological load causing ecological crises which result in dead water is here transformed by movement-nurtured biological processes into life-filled water.

draw an uncontrolled wiggly line it has no form, other than, as we discussed earlier, any organization brought by your bodily movement imprinted on it.

If you breathe with the line — breathing out, breathing in, accelerating, decelerating — the curves have quite a different quality.

This is the basis of Celtic and other 'knot work' design — bringing breathing life into line. By controlling the visual 'breath' they could touch upon the same archetypal focus that the actual breath does in speech. For the sounds of speech are not random; they are related both orally and physiologically to how we use our breath to create mood and meaning without words. In this sense, we can inter-

We can look at the Chinese tradition of feng-shui as conversation between matter and cosmos: planetary qualities given earthly embodiment and in appropriate relationships to each other were considered to be health giving. Curvilinear lines bring where they come from, where they go to and what they pass through into receptive conversation with each other. Straight lines with their insensitive, imposed and alien nature obstruct and destroy this harmonious, life-giving force.

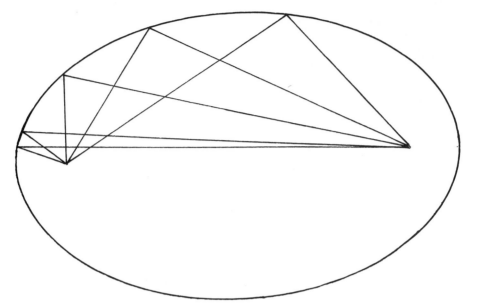

pret Gothic architecture with its distinctively active curves as speech without words, as spirit-meaning written into architecture. This is only one aspect, of course. It took most of a lifetime of occult study to master these secrets in Gothic times. Nowadays we have lost all connection with that secret wisdom, but we must find our access to the effects of shapes by cultivating a consciousness of what we actually experience in differently shaped spaces: spaces which welcome, exclude, are tense, relaxing, dominating, or allow us to feel we are free individuals.

To understand their nature we can also make an intellectual connection with curves through mathematics and science. Like a circle, an ellipse is formed with a fixed length of string, but it has two foci. It is therefore a form which has both one-ness and two-ness in it. It makes a good shape for a hierarchical room with democratic preten-

An ellipse is anchored by an additive formula (a + b = c) to two points. To make one you need a bit of string with a nail at each end; keeping the string taut move the pencil around the nails.

sions — conference tables are often this shape. However, where the hierarchy is very strong they can be trapezium-shaped. Such conference tables are made and are of course very practical; everyone can see and is focused towards the principal character. Auditoriums are often variations on this shape, in section as well as plan. These are shapes for hierarchical relationships. If we wish to be more egalitarian, for instance to turn a 'lecture' into a discussion, we pull our chairs into a circle.

Scientific observations throw another light upon straight lines and curves. 'Design lines' — straight lines to destinations — are an abstraction. Animals walk in curves. So do people. We do not change direction in exact abrupt steps as does the computer screen but all in the fluid process of walking. Studies of moving fluids show that they always seek curvilinear forms. No river is straight unless it is confined by uncompromising geology or canalization. The slower a river flows, the more pronounced is its meandering form.

The meander isn't made up of bits of circles, but a live relationship between that which is accelerating into curve and that which is decelerating into straightness and eventually into a curve the other way. Accelerating-decelerating curves are always therefore in a ten-

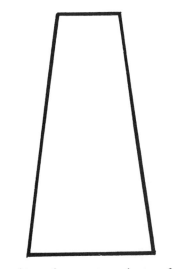

The hierarchy-stressing shape of the trapezium.

The slower a river flows, the more pronounced is its meandering form.

The meander is not made up of regular circles.

While curves often have a welcome soften-ing, harmonious effect on buildings, alien added curves jar more than any straight lines.

sion between straightness and curve. Segments of a circle are, by contrast, static.

From time to time builders ask me what the radius of the curve of a particular window head is. I never used to know. I just draw the shape that feels right to me. Now I real-ize that these curves are *not* the arc of a circle. Recently a new fashion has developed in British bungalows to use arches, often using arcs of cir-cles which have no relationship to any other forms.

If you make (or even just trace your pencil point over) an acceler-

Arcs of circles without relationship to other forms.

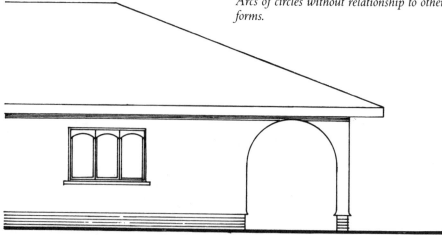

ating curve you can get a feeling for the latent strength that is built up by the accelerating concentration which the line shapes. If you can develop this feeling for life in curves like this, you can put flat planes like plasterboard together and imply curves.

This is all about *spaces*, but forms can also have powerful effects. We don't have to travel to Moscow to experience how rows of rectangular giants chillingly oppress the freedom of the individual. In the Soviet Union, as indeed all over the world, the intention was to give people homes. Bureaucracies tend to approach people as numerical statistics, but a short step from treating them as material objects. There is also a tendency to assume a dependent relationship between individuals and the accommodation-providing institution. The tower blocks that resulted were appropriate forms for such a philosophy. They were of course a quite unconscious choice of form — and because it was unconscious, the results were inevitable.

It isn't just the boxes, their huge size and the absence of distinction of individual homes that are oppressive, but also their repetition. An abstract idea can be repeated endlessly until it comes into relation with reality, then it has to be modified. In every seed there is, more or less, a pure plant, an archetype. As the plant grows the individuality of its surroundings — soil, climate and so on — causes it to modify this archetype. If you look closely, two adjacent clumps of grass are different. Trees can be so different that we can recognize them as individualities. Even development in time is a metamorphic process: young and mature and early and late leaves have different shapes.

When we build objects that do not evolve, we deny this life process and this response to surroundings. We impose dead ideas, often ideas which are not even modified by materials (in the way that clouds often form the shape of a landscape, itself made up of a metamorphic series of forms).

Commonly, built forms are

Curved enclosure made of straight-line material such as planes of plasterboard.

Accelerating curve built out of straight lines.

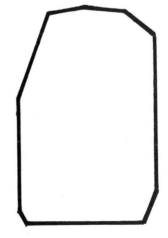

Repetition without metamorphosis.

repeated. When you repeat things and the spaces between them, you start to make a rhythm. If neither the spaces nor the objects change, the rhythm becomes boring, dead and compelling. To leave the listener or observer free both objects and spaces need to evolve, to respond to one another. This way the active listener participates in the rhythms of life: growth, decline, substantiation, enrichment, inversion and so on. Interwoven with the rhythmic metre of the body — heart and breath, and through the ear, with its direct link to the soul — this is one essential that distinguishes music as art from music as manipulator. Beat without metamorphosis, merely reflecting bodily rhythms, floods the soul with bodily desire-emotions, overriding individual judgement. It is no empty coincidence that armies march to the drum.

If, without variation, we repeat a window shape, we take no account of its different relationship and functions outside or inside the building. The vertebrae of the spine each carry a slightly different load and accept a slightly different movement. They are not identical. Each window likewise has an individual set of requirements to fulfil — unless we are just providing containers for people, albeit elaborate ones. It may be ridiculous to make every window different just for the sake of being different, but it is even more so to make every one the same just for the sake of being the same, or to shape them just to impose an elevational pattern.

Repetition is the basis of rhythm. It can bring an anchoring structure, but organization by repetition is organization by the imposition of lifeless systems; think of an avenue with identically pruned trees.

Organization by metamorphosis arises out of living processes; life brings the disparate into unity.

It is not difficult to design powerful forms especially if they are large. What *is* difficult is to bring life to them, making them inviting neighbours to live with. Dynamic shapes can be unsettling, especially two-dimensional simplification emphasized by rectangular plans. These are the product of *shape* rather than *form* consciousness.

If you make sculpture you can approach the form from the outside, imposing shape-sections upon it, or you can work like Giacometti, with the energy bursting from the centre out. You can also push things around so that one form changes into another. You can design build-

Like strong colours, repetition leaves one unfree. It is hard not to give yourself over to repetitive music. Rhythm brings the first transformative influence of life; metamorphose carries it further.

ings in these three ways: you can impose shapes from the outside creating three-dimensional shapes but never forms, or allow non-rectangular plans and sections (which describe spacial experience) to create non-rectangular forms. You can develop these further by means of a clay model, which it is easy to push around to try out variations.

Of course you can make anything in clay, so you often need to follow this up with a constructional model. There is also a risk of thinking about a project more in terms of forms than of spaces so I like to keep both clay model and sectional room- and place-sketches going at the same time. Once one works from the inside out, even simple plans with a simply ridged roof can produce a variety of forms depending on where the ridge is, whether the eaves or ridge are horizontal and whether the pitch is constant.

This is even more the case when

Even a simply roofed plan can give rise to a variety of forms.

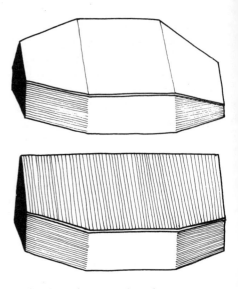

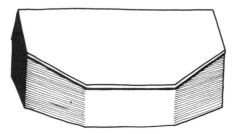

we consider conic sections. The form of the Steiner kindergarten I am building at the moment is based on two cones rising off square plans; but the planar sides are not at right angles, are different sizes, and the two principal units, together with the simple roof over the rest of it, all are in conversation with one another.

All these sort of forms are more self-contained than simple rectan-

guloids. They tend to be stronger as objects and therefore to be harder to make into satisfactory boundaries of space. The most difficult objects of all are spheres. I have not yet seen satisfactory external enclosures created between spherical buildings. However, personally freeing their

The outsides of circular space — circular forms — are difficult to bring together to make places *with. It is possible but much harder than with straight-line forms.*

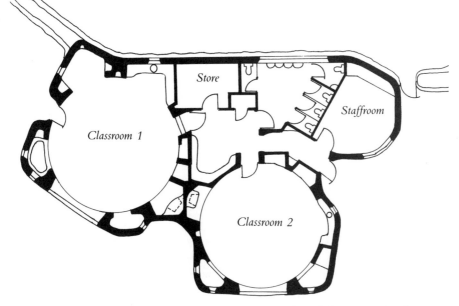

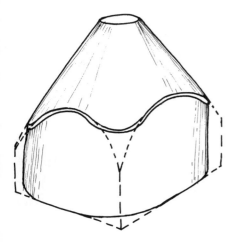

Above *Plan of the Steiner kindergarten.*

Left *The kindergarten is made up of social (circular) and individual-play (corner) shapes on plan, making, in form, cones on square bases.*

Below left *The interaction between cone and (rounded) square gives the roof eaves this undulating curve.*

Below *These curves can become gestures of entry and outlook.*

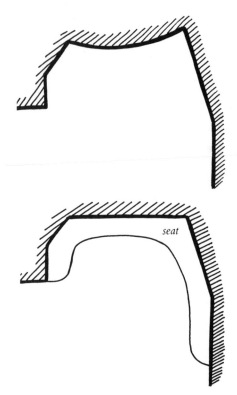

interior spaces, egg-shaped 'living-pods' or geodesic domes clustered together can achieve no better than suburban spacial relationships — objects with left-over space between them.

Non-rectangular spaces are more life enhancing both inside and out-side. But they are much much harder to work with as enclosures of place — harder, but possible. The issue is not between rectangles and non-rectangles: why do something differently from the normal way unless there is a good reason? The issue is between living and lifeless forms and spaces, life-renewing and life-sapping environments. Cutting off corners, non-rectangularizing or curvilinearizing shapes won't do more than make buildings look curious unless we work seriously with the question, what shape-language is appropriate for human — and therefore spiritual — needs?

The same problem can occur even inside buildings with backs of some space-enclosing forms intruding upon other spaces. In this project the plan of the structural walls (top) needs to be transformed from a repelling to a welcoming gesture (above).

Non-rectangular forms are less harsh and oppressive than rectangular ones — just imagine the experience of walking around a right-angle cornered building or one at 105°.

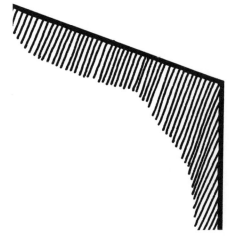

8

Design as a Listening Process: Creating Places with Users and Builders

We have considered the proportions of space and how to make spaces appropriate to how they will be used. The right spaces, however, are only a beginning. Unless we are architects, we don't pass from one space to another; we go from place to place. But how do places come into being?

Traditionally places grew slowly, each additional building, each reinforcement and evolution to the sense of place meeting a need. First perhaps there was a ford over a river, a crossroads, perhaps a few houses, then a bridge was built. Then came the tavern, blacksmith, village square, a little way away a sawmill, more houses squeezed between hillside and river, then a fountain-trough for horses to drink, a chapel and so on. The place, made up of different sub-places, has developed a definite and individual quality. Different political, geographical, economic and cultural situations influenced the pattern and sequence of growth, producing very varied forms of village. None the less the principle underlying organic growth is that the physical form grew out of *activities*, out of the meeting of users and environment.

It doesn't usually happen like that these days.

The process of design and building is now much more rapid. There is little room for evolution, little room to correct things even if they become obvious before the building is finished. The owners state their needs — which may or may not have anything to do with the needs of the place. Buildings and the spaces around and between them often are little more than the material enclosers of quantitative space allocated to activities — so many cubic metres for this or that.

I am lucky in that these sort of people don't normally come to me. If they do it soon becomes obvious that we do not share the same values and cannot work together and, if they don't realize it themselves, I have to tell them that I can't work with them.

When someone engages a professional adviser to undertake a project, a relationship of specialist and client is established. Normally the client states specific functional objectives and the specialist is free to find the most appropriate form to satisfy these. In architecture this freedom can be interpreted as the opportunity to implant an individualistic expression into the design. A relationship of delegation, where the architect's job is isolated from the project initiator, just to 'get on with it', encourages this designer-ego-trip approach.

I can't work this way: my work *depends absolutely* on my client. I don't feel I have the right to make decisions if they do not have clients' support. After all, it is they who will bear the huge expense of building, so I have absolutely no right to just indulge my own whims. On the other hand, I hope that I have come further in my ability to distinguish personal preferences from deeper and more widely experienced responses to qualities of environment. As I know that environment has definite harmful or health-giving effects upon occupants and users, my first responsibility therefore must be to these people. Many of them I may never meet; many may not even have been born yet. Consequently I cannot just bend in whatever direction a client may ask and that is why it is so vital that we both share the same values.

For many smaller projects, my clients and users are one and the same person, which makes things easier. Their needs, however, are frequently obscured by rigid ideas about what they want built.

The process of design for a small project, like a one-family house, a craft workshop, a café or farm building, usually goes like this: the owner-users and I sit down together for a mutual design session that I *know* — though they don't usually believe me — will probably last five, possibly eight, hours. We have a lot of semi-transparent paper and a scale drawing of the site or building to be converted, and we will have already walked around, discussed and observed possibilities and limitations on site. Invariably they know what they want. They have seen things elsewhere or in advertisements. They want these things to be put together and that is my job. Is it? Are these frozen concepts what they need, even what they really *want*?

I try to open up the implications of what they ask for. We identify the conflicting requirements, the priorities. We work towards the *qualities* they are looking for, qualities that had become frozen in their minds into objects, adjectives into nouns. Now, with these qualities, we get close to what they *really* want.

I describe this with words, but actually most of it is done with a pencil. All the time I am trying to describe — sometimes in words which evoke examples we share experience of, but more often by sketching, the *limitations* of every suggestion and its *possibilities*. The drawing process, therefore, alternates continually between plan laid over plan which show how things are arranged, sections which show the space in which we live, the views, light, etc., and sketches — mostly interiors of rooms or of external spaces, of how it will *appear*.

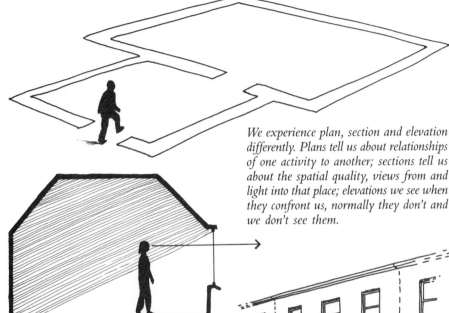

We experience plan, section and elevation differently. Plans tell us about relationships of one activity to another; sections tell us about the spatial quality, views from and light into that place; elevations we see when they confront us, normally they don't and we don't see them.

My own priorities in design are place, relationships and sequential experience. It is difficult to give an order to a process which jumps continually from one mode to another, but walking the site, plans, sections, clay models (form), sketches (mood and sequence) and card models (space) are all involved. Elevations, my main concern as a student, are only the servant of the experience I wish to create.

From time to time we need to get up to chalk out spaces on the floor — are they big enough? Too big? We measure familiar spaces, move furniture around to show how big such and such a room will be, and so on. If we are near the site we can put sticks in the ground to show where the buildings will be, what view there will be past them, how they enclose outdoor space, articulate pathways. Or, in existing buildings, we can chalk out shapes on the walls for windows, lay planks across the floor for walls; we can stand and try to imagine the view through a wall that one day won't be there.

It is difficult to project the imagination beyond what we can see. There comes a time when we need

97

With clay, the form of buildings and their larger external spaces can be rapidly developed and continually altered. Drawings can be measured off the model, and themselves adapted as required by dimensional, constructional and any other requirements which are too imprecise in the clay model.

A working card model simulates the constructional limitations of carpentry. You can make any sort of curved surface in clay, but in cardboard you must make rafters and other shaping pieces just as though you were really building the form.

All models however have the limitation that they are so frequently viewed from above that it is hard to remember that in reality we will get a much more limited experience of form and space; minor elements like a climber-overgrown trellis fence hardly show on the model but in reality enclose the space we see. Then there can be seasonal barriers — a few bare twigs in winter but a wall of leaf in summer. Similarly, volume and speed of traffic can turn a dimensionally inconsequential open space into a major division and focus attention on whichever side of this invisible barrier we happen to be.

the abstraction of paper to think beyond the confines of the present. There are other things only paper will tell us, organization of the building through barriers to sight, particularly floors. We can also check things quickly: perhaps now we have to cut out the shapes of a few special pieces of furniture — grandfather's table perhaps — will it fit? Will it fit through the doors?

We are now getting near to a mutually agreed design but it is never too late to step to the side and try a *completely different* layout. Most projects develop several families of plan before we decide on just one.

At the end of this session we are all rather tired! But what we have done is to design something *together*. It isn't my design; nor is it anyone else's. Each of us has brought something individual, from our own unique biographical stream of development, but it has been *given* to the whole. The design has arisen out of the conversation *between* us. The result is much better than I could possibly have achieved on my own. That is why I say that I depend upon the client, and why I must be confident that we share fundamental, usually unspoken values — even if outwardly they live a completely different style of life.

At the heart of the process is listening — one of the most difficult things in the world to do! For to listen you must put your thoughts and opinions aside and listen acutely to what the other person means but is perhaps unable to say properly — even if you don't like what you hear. *Listening* is the opposite of *argument*. Arguments polarize people's positions until they cannot listen, but they do not produce anything. I have never convinced anybody (nor been converted) by an argument whether I won it or lost it. But out of *listening* the right questions arise. Conversational design depends upon the right questions.

I find this is an obvious way to go about design. In fact I find it impossible if I have no user to talk with. I just can't make any headway with competitions, for instance, for just this reason. In isolation I can't ask anyone questions. I have no way forward — only my own ego-trip ideas and these have no way even of being adapted to meet anyone else's needs.

Conversational design is easy to practice in pairs. One person takes the user's part and answers straight from the heart. The other — 'the architect' — must hold back his or her own preferences and only offer 'what if. . .? or 'do you mean like this?' suggestions. The client speaks with words, the architect supports everything he or she says with

sketches — the less words and the more pencil the better. To make the sketches meaningful, you can measure the sizes of rooms and chalk them on the floor.

Conversational design is about finding appropriate *qualities*, adjectives (even if the user at first asks for objects, nouns), and, with imagination, sensitivity and experience, giving them possible forms. It is *not* about imaginatively thinking up qualities and combining them to make individual statements. That is how I work with owner-users who comprise only two or three people and buildings that are reasonably small. It is much harder with larger groups of people and buildings.

Large projects tend to be too complicated to design on the spot. Large groups of people take a long time to come to agreement on any one thing. There is also the problem that some people come to one meeting, agree something, and then other people at the next meeting want something different! A major difficulty is that people are often too busy to find the time to take part in the design process. Many institutions have got used to delegating design and just waiting for the results to be provided — yet when they start to experience design as a living process, the social, environmental and practical issues involved can become fired with a new inspiring imagination. People can begin to realize: 'This is what it *could* be, what it *should* be. Together we can make it possible.'

Some aspects of a project need the involvement of everybody who has an interest in it, some concern only more specialized users: caretakers, cooks, teachers and so on. I therefore try to sort out which aspects of design we can all work together on and which need only a few of us. I also make the point that anyone who is *committed* to the project can take part as much as they wish, but commitment means committing *time*. If they are too busy to take part in design but only look at and object to finished proposals, we will maximize only the negative aspects of participation. I want the process to be positive.

The first thing we need to decide is what are the organizing principles? Whether people are sufficiently conscious to voice them or not, these principles are of a qualitative nature. There may for instance be a lot of activities which need to be accommodated; some, for practical reasons, need to go next to each other; some, for environmental reasons, best suit certain parts of the site; most can be anywhere; but each activity has a different quality. Each quality has a 'colour', a mood which distinguishes, say, an exclusive cellar restaurant from a McDonalds hamburger bar or a library which reminds us of a Rembrandt from one like a Mondrian. When we put them together in one combination or another we can create courtyards, alleys, places of spirit or of non-identity. Medieval towns had streets of shoemakers, saddlers, jewellers, cloth merchants, hay markets. I am not talking about *zoning*: a street you can walk through in a few minutes — a zone you must drive through.

As an example I was recently advising a Steiner school in its future expansion plans. The site is a long strip along a hillside — a street. The school is made up of a number of activities; each activity 'colours' the space it opens on to. We can therefore create instead of a road with buildings beside it a walking *street* with courtyards each with its own 'colour', its own local spirit.

Once we have talked a bit about activities and got beyond any fixed conceptions of such-and-such *has* to go here or *cannot* go here, we can walk the site and begin to visualize what activities suit which places, how would the environment be altered? How would the 'colour' of the activity benefit from and *contribute to* this particular place?

Once we have developed this 'mood picture' of the site we can go indoors and gather round a table with a large site plan, scale 1:50 if possible. We also need some coloured paper — one colour for each family of activities — and a large quantity of cheap semi-transparent paper (e.g. greaseproof paper). We can cut the coloured paper to the size and approximate optimum rectangle for each main room such as classrooms, kitchens or workshops. The rooms may not be that shape, of course, but we can rapidly check whether they are about the right size. The thin paper is for us to draw overlays of where buildings might be. As we go I can sketch the possible appearance of what we have planned. At the end of the session we can hope to have a plan of what we all would like to see. We can now take the chosen overlays off and go through the process in the probable sequence of building phases. Does it grow *properly*? If the project never completes its final phase, will it still be good?

After this, back out to site to walk around — things aren't always as one remembers them. More sketches here, standing in particular places, of the view as it might become, of the passageways, portals, steps, turns and changes of textures that articulate space.

Prior to this exercise it is sometimes worth trying to develop a feeling for environment appropriate to particular activities in *colour*. We try to paint the colour mood of a place or sequence without copying the actual colours it will be built in and without drawing any objects. It is only in colour. Because it requires us to leave behind all our conventional (and blinkering) anchors, this a difficult exercise. Although the resulting paintings may well be disappointing, the experience of doing them helps to identify the soul-mood which is at the vital heart of the project.

It is usually time now for me to go away and develop the design. As I do so, of course, it becomes obvious that I have to make changes here and there, perhaps even quite major changes — some things just won't fit in without excessive cost, such as loss of open space or overshading. Here I need to have access to a small group who can make rapid decisions so that I can keep on working.

It is useful at this stage to use a clay model as a design tool, for clay can be pushed around whenever you wish to see what alternatives look like. A card model, however

quickly made, is far too inflexible and tends to fix one's thoughts. A clay model would be excellent for the users to manipulate, but — apart from student group projects — I have found it too difficult to arrange.

Now is the time for another large meeting with more detailed, but still rather flexible, proposals and a lot of sketches — for it is sketches that people understand. When we have moved things around a bit, walked them on site and (hopefully) agreed in principle, we can go on to discuss specific areas with the specialist users.

Not every group wants this sort of involvement: it is after all very time-demanding. The more usual way is for one person to brief the architect, answer any queries that arise during design, and then a committee convenes to assess, and criticize, the proposals. This tends to maximize the negative aspects of social interaction. If we wish to develop the positive aspects, we need a different structure to these relationships and a more living process.

A fundamental part of this process is listening to the meeting of the idea, the 'colour' it will give a project and the environment which is already there: listening to the spirit of the project coming into being, that invisible ideal which is so much more than the sum of its individual components and functional requirements. It is possible to cultivate this ability by looking at places one knows and thinking, what activity — and therefore what qualities of building or landscaping — would *add* to this place, help it

grow to be even better than it is at the moment?

Of course, we have to be *very* objectively critical of our own thoughts, otherwise it is possible to justify all the desecrations of landscape and townscape that we see around us. If we approach places like this we are thinking no longer in terms of places as opportunities for the imposition of great ideas, but of developing what is already there. In the same way that we think of conversion of existing buildings, we can think of conversion of places.

Sooner or later these buildings we have designed actually get built. Now what were previously thoughts recorded on flat paper, or — at best — seen from above and outside as models, become actual spaces. As they grow to completion, all sorts of things begin to show up. What previously we could only imperfectly imagine can now be seen! Users can start to see the ways they will use their rooms — and also the mistakes I have made! We can also see a lot of things we could previously only guess at: views past and sunlight through trees may only require the cutting of selected branches. These just can't be anticipated or even identified on paper. Nor until we stand in the half-built building can we see the views from new floor levels or the ways various architectural elements meet, like walls, ceiling planes and doors. To bring these meetings into conversation we can chalk out shapes, try them out with battens lightly nailed in place. This way of listening to how places want to develop is something one can practise. It is interesting to go onto unfinished sites without having seen the architect's plans and think, how would I develop these forms, spaces and

Hand construction allows builders to become artistically involved in their work. It allows on-the-spot design so that, as here, the ceiling shape can metamorphose from one situation to another in a way that paper design can, at best, freeze into a lifeless diagram. If labour costs can be freed from conventional time = money formulae by more creativity-enhancing contractual arrangements, the costs need not be penal. This school, built by volunteers and therefore with free labour, cost approximately a seventh of estimated contract price.

meetings of elements that are now emerging?

As building progresses we can actually experience entry and movement sequences, the ways views are focused, sunlight penetration at particular times of day and year and many other things we had no way of visualizing before. We have the opportunity to emphasize or moderate these experiences.

Once we think of the creation of places as a *process* it is obvious that if places are to develop in a healthy way, every stage needs not only to be healthy but to add something beneficial to whatever has gone before. We can start to think not only of architecture as developing the currents of people's and places' needs but of detailed constructional design, then building construction, then occupancy and use, further developing this process.

Each stage builds on the one before and opens itself to the one after. It depends very strongly on what the preceding stages have provided as a starting point. Architecture is just that — the bringing into relationship of the background context and the life of future users. If either of these gets left out, it ceases to be relevant, opening itself to resentments. It loses its place in the stream of time. If it can weave together what has come before — the environmental context — and what *will* come — the users — and through the process of designing and making raise the ingredients artistically, it can find in a *new, conscious* and *relevant* way the organic process on which the evolution of places depends.

Overleaf *Helping a place come into being: at every stage, the place needs to be both complete — have a quality of belonging, of eternity — and also allow opportunities for future growth. The present, grown out of the past, needs to be both complete in itself and open to the future. I have designed too many buildings shut to the future; now, more concerned with place and the life within it, I try not to, for living, growing places are founded on the meeting between activities, users and environment. This is a farm course centre for city children and handicapped people.*

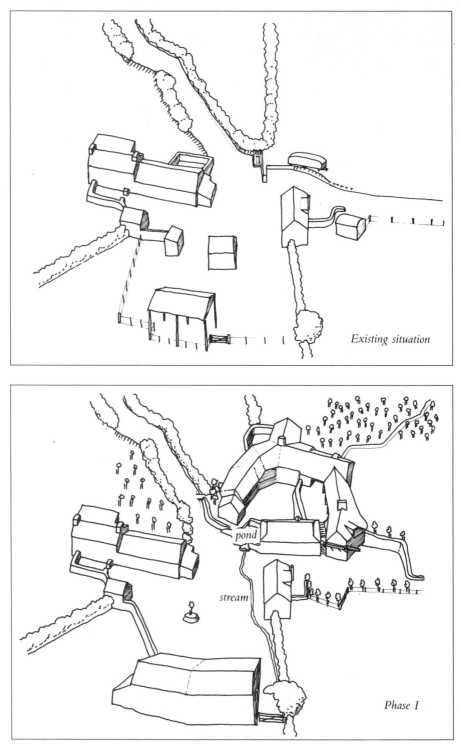

Existing situation

pond

stream

Phase I

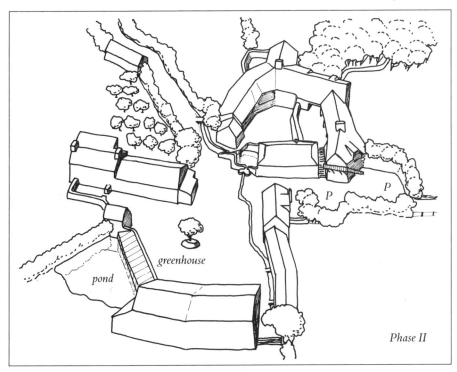

greenhouse

pond

Phase II

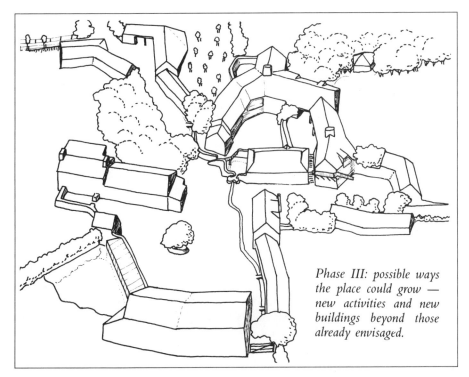

Phase III: possible ways the place could grow — new activities and new buildings beyond those already envisaged.

9

Ensouling Buildings

If you stand in front of, or go into, a new building nowadays the usual experience is one of emptiness. It waits for someone to come along and give it love, cosiness, individuality, to put curtains in the windows, flowers on the balcony, life in the rooms. And so it should! Until that someone comes along, however, many buildings are lifeless. They offer nothing other than constraints and architectural qualities — space, light and so on — to build upon, to work together with. Such buildings have not yet started the process of being ensouled.

So what is this process of giving a building a soul?

Soul can incarnate progressively into a building as it progressively gains substance from wish, through idea, planning, constructional design, building and occupation. Each stage develops, deepens and extends that which had come before. They are not stages which alternate from aesthetic to practical but, with these aspects inseparable throughout, are stages of continuous process of incarnation into sub-stance until we architects complete our task, leaving a shell for life which will continue to grow.

It is conspicuous that buildings which have been designed and built without care, or where their tenancy and management structure discourages tenant care, replacing it with dependency on a faceless or exploitive owner, rapidly deteriorate into slums. A generation ago, slums used to mean buildings with physical deficiencies. Today's slums are buildings from which care is absent.

Old buildings are rarely just museums of a particular period of history. They have physical elements from many dates right up to the present and they have the imprint — both visibly and invisibly — of the many occupants, lifestyles and values that inhabited them. When this has been a harmonious progression, the new built upon the past rather than brutally pushing it aside, ripping it out and trampling it, these buildings have charm and appeal.

People choose old buildings to live in. Of course people also choose to live in other sorts of buildings, but for other reasons. What I have described for buildings also applies to landscapes: everything new that we build will be set in a landscape or townscape that already exists and which has been made up by a long historical process. What we tend to call *sites* are already *places*, places to which their histories have given soul and spirit.

The soul of a place is the intangible feeling — made up of so many things — that it conveys. It is for instance sleepy, smells of pine trees, is friendly, airy, quiet, its roads and paths do not hurry but turn slightly so that everything can always be seen anew each time you pass. Upon this composite of sensory experiences, reinforced by historical associations ('under this clock is where couples always met, even my grandparents', 'this is where the great ships were built', etc.) we begin to feel that there is something special about this place, unique, living and evolving, but enduring beyond minor change. It is a being in itself. I call this the spirit of a place. Every place should have a spirit; indeed, unless it has been destroyed by brutal unresponsive actions, every place does.

Children know every corner of the little piece of land they play on. It gives them happiness and health forces they will carry into later life. To the small child it is a whole world, every part an individuality and large in area. Revisit it as an adult and it seems tiny. Revisit it as a site manager and 'here we can stack the concrete units, here the reinforcing steel; we need only to level the site first'. Nowadays so much land is *used*, so little *appreciated*.

When we think of projects these days — urban redevelopment, housing estates, motorway junctions, oil terminals, airports — how many places with a special, unique, valuable and health-giving spirit cease to exist? Whenever we build something new we have a responsibility to this spirit of place. A responsibility to add to it. To the Ancient Greeks the sense of these beings was so strong that in particular places they could say 'here lives the god'. They then enclosed and strengthened this being with a temple.

Today our buildings serve different functions — inside and outside ones. *Inside* is to house an idea, say a clinic, a shop, a home. *Outside* they bound, articulate, focus or alter an external space, adding to or detracting from what is already there, the spirit of place. Many outside spaces serve both functions — an 'idea' function (like a meditation garden, private courtyard or car park) and a 'response to place' function.

Because the inside space, activities and qualities of a building and the outside surfaces and appearance are interrelated, the whole building and all the activities it generates need to be involved in this great conversation. The conversation between idea, usage and place, between what will be and what already was. Between physical substance — the materialization of the idea — and invisible spirit of place — the spirit brought into being by the physical substance of the surroundings. This is a fundamental responsibility in any architectural action. At first sight it might seem too big a responsibility to cope with, but I don't approach it like that. What I do first is try to listen to the place, listen to the idea and find ways in which they are at least compatible. At best they can symbiotically reinforce each other. Then I try to design a building which has the appropriate qualities. In a landscape, this often requires a building which is as small in scale as possible. This can be achieved for instance with low eaves, preferably below eye-level, or by tucking the building into the landform or placing it so as to extend or turn the lines of hedgerows or landform features.

Townscape situations have different requirements. Perhaps the critical issue might be to find the right scale and intensity of visible activity while at the same time minimizing adverse effects like shading or noise, especially in the more sensitive neighbouring zones. Enclosure, compression, openness, sunlight, activity, vegetation, airways, acoustic textures (like plank pavements) and so on are all things that a place can ask for, that can bring a benefit to what already is. Light (including colour), life (especially vegetation), air quality and noise reduction (especially mechanical acoustic noise, but also sensory, especially visual 'noise') and spatial variety, meaningful to the soul, are likely to be amongst the critical elements.

The building itself needs appropriate qualities which both add to how it looks from outside and colour the *activities* within and around it (these activities may well have a greater impact on the surroundings than the building itself will!) It needs a meaningful choice of materials. Traditionally materials found in the surroundings were raised artistically to become buildings. Today we are free to use anything. But to fit, the materials need to feel right for the place. When I go to a new site, I can feel for example, a stone or a wooden building

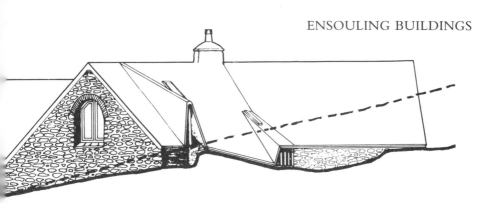

would be right here; or perhaps for this use, the building needs to assert its purposes a little more and should be rendered block; here it should be more urban in character — brick perhaps?

A building needs forms and shapes — outlines — roof and eaves lines which relate to (not necessarily copy) or perhaps contrast the surroundings. These, combined with plan shape, create the appropriate gestures: of welcome, of privacy, of activity, of repose. These in turn are part of the experience of approaching and entering a building. Roads, paths, boundaries (such as fences or woodland edge) and topographic features (such as the junction of sloping and level land) tie a building into the landscape. The 'keyline' system of erosion control and fertility building is generated from the 'key' topographic

Existing features, such as roads and hedgerows, not only tie a building into a place; they can suggest, even ask that it be there.

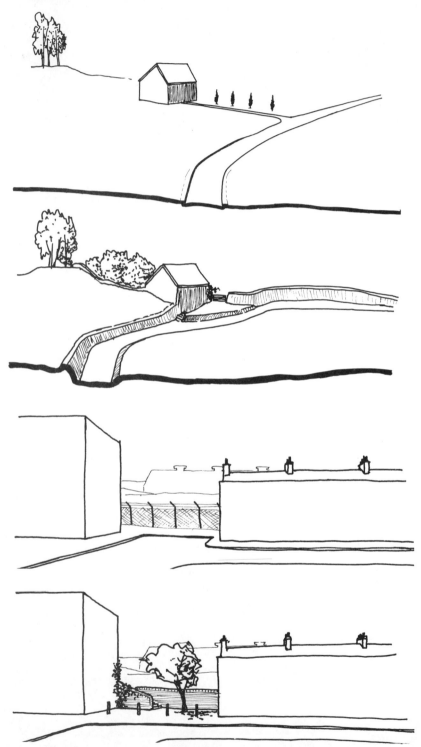

meeting point between steep and flattening slope.[1] The placing of woodland, roads and fences is critical. It is no coincidence that these can make a development belong to a place or, if unconsidered, assault it spiritually and ecologically. They are all features which are either already there and anchor that which is new, or are implied by the relationship between the existing situation and the new building.

I notice that, quite unconsciously, I often prefer to locate buildings on the edge of a site where there is something, a wall perhaps or meeting point of different qualities of place, out of which the building can grow. There will also be more open space left on site to do something attractive with, and not just bits of left-over space all around the building.

In built-up areas where open space and sunlight are at such a premium, buildings placed to dominate the site are spatially — and in this commercial world monetarily an extravagance we can rarely afford. Buildings sited to give priority to the place they bound make better environmental and economic sense.

As one approaches a building there is a moment when you come to be aware of the influence of its *activities*. This is a threshold. It is the place for a bridge or archway, either built, formed of trees meeting overhead or implied by buildings compressing and focusing space. There are other ways of giving emphasis to this threshold, like using a turn in the path around a building corner, a group of trees or slope of land, a change in ground surface such as from long to mown grass or gravel to brick paving. Gates and

Even in a small country like Britain much land is wasted. They are the forgotten spaces: behind the garden shed, the other side of the factory fence, behind the dustbins. Some are miniature wildlife havens but most are just places of squalor. Children need such hidden places, but not the ugliness, the excreta of society. Turning our backs on things, pretending they do not exist, is quite the opposite from not interfering with a place. Britain can justly be proud of its tradition of backs to houses: many countries have fronts all round. But a house back is the front to more private and less conformist activities. It is the lifeless backs that I am concerned about — so much land in despair that could be home to life.

[1] This system, developed by P.A. Yeomans for Australian climatic conditions, has been widely and successfully used to reclaim and improve dryland prone to infrequent but destructively heavy rain.

steps are traditional threshold markers.

If we are to bring anything new to a place and make it better, not worse, that new thing must have an artistic quality. Art starts when inspiration struggles with the constraints of matter. When the painter paints, any pre-formed idea has to give way to what is developing on the canvas; matter and spirit become interwoven into a single whole. The idea on its own existed outside the sphere of earthly reality or life — the painting process gives it reality and life.

This process applies as much in architecture as in any other art. First someone perceives a need, sometimes a set of needs; then comes the idea — how to satisfy this need; then an architectural concept; then a building plan, constructional design, a period of building longer and using more energy and money than the previous planning stages; then use — even longer and where the building affects the occupants and users every day!

Conventionally, artistic design stops at or only a little beyond the planning stage. But most of the work is still to come. If any product is to be artistic, the people who make it need to be involved in the artistic process. Of course, builders have not necessarily gone through the same process of developing their aesthetic sensitivities as have architects, but there are other ways to look at it.

It is often said: 'What is wrong with this region is that there is no overall planning!' We live daily in localized experience, all influenced by a regional structure. Our local world is the victim or beneficiary of mega-decisions: after Regional Planning comes District Planning (we begin to *see* the consequences here) — then architecture. Architects tend to agree how important this is. Then the textures, loving craftsmanship (or otherwise) with which things are built, then fur- and building maintenance. Then *homemaking* — both at home and at work — perhaps the most important stage of all that makes places welcoming and our lives a pleasure — or not! If left out, it undoes all the good built up so far.

Generations of care and life give old buildings their charm; lack of it turns them rapidly into slums. The *architectural* qualities have but a small part to play in this spirit that grows up in places. I am reminded of this every time I see an attractive but empty holiday home. Yet it is everything that has gone before that influences whether places will be loved and cared for or resented and abused. Only for a century or so has this whole process been compartmentalized so that aesthetics is restricted to the architectural stage. Yet great ideas badly, carelessly, lovelessly built are awful to live in! Many qualities depend upon *how* they are made.

Many of the finer qualities of a space — the complexity of meeting forms and planes, the metamorphosis of one shape, form, space into another, the effects of natural and artificial light — can only be approximately and inadequately anticipated. They must be made.

Dead straight lines are so *dead*. To give them life they need to be not wobbly, random or weak but made

The attitude, artistic and care-filled or otherwise, with which a building is built makes all the difference to the end product.

with a feeling hand. Made. This is the sphere where *only* the building workers can make or break a building. When you make things with your own hands you just can't make *satisfactorily* the same form in different materials. It feels different, needs a different structure and form.

Making and building things is the stage at which idea meets material. They can either compromise each other or, through their fusion, reach a higher level. Sculpture in the mind is pointless. Without art, stone fresh from the quarry is little more than a pile of broken rock. It *is*, however, a *little* more than just a pile because each material already has something in it waiting to find an appropriate place and

form. Not every stone has Michelangelo's David in it, but every stone has a quality of 'stoneness'. The violence of the quarry leaves it with sharp split surfaces, but the quality of enduring rock *can* be refound.

All materials have individual qualities. Wood is warm, it has a life to it even though the tree is long felled; brick still has, to touch and to the eye, some of the warmth of the brick kiln; steel is hard, cold, bearing the impress of the hard, powerful industrial machines that rolled or pressed it; plastic has something of the alien molecular technology of which it is made, standing outside the realm of life and, like reinforced concrete, bound by no visible structural rules. It is out of these qualities that materials speak. It is hard to make a cold-feeling room out of unpainted wood, hard to make a warm, soft, approachable room out of concrete.

Materials are the raw ingredients of art. But already they affect our emotions. Mediocre architecture on a scale that is not oppressive is really quite pleasant in timber or a well-chosen brick but a disaster in concrete or asbestos-cement panel.

On the whole, people don't look at architecture, nor at materials. They breathe it in. It provides an atmosphere, not a pictorial scene. When you look at a photograph of an attractive place you notice how much of the picture is ground surface. Our field of vision usually includes more ground than sky. Our feet walk on it. The materials of the ground surface are *at least as important* as those of the walls.

Beyond individual personal preferences we respond to the his-

Above *Much more than the architecture, it is the materials and play of light upon them that make the atmosphere of this place.*

Below *Ground texture and vegetation are often the most expendable items of an architectural budget. Yet they can be the most economical and effective elements in making a place.*

tory and 'being' of the material printed into its appearance. Our feelings are not random but relate to how appropriate this 'being' is to our needs of soul. They also are closely interwoven with the effects that the material has on the body.

Biologically and emotionally metal, reinforced concrete and plastic are not good materials to live within, but wood is — very much so. Nowadays when we think of wood we picture not the curving branches and forks chosen by the shipwrights and early framed–house builders but machine-extruded strips. These lend themselves well to planes but poorly to curves — the opposite of brick where curves can give such strength you can't push over a tall narrow wall. In wood, I usually make curved gestures out of straight lines. Three-facet arches, polygonal spaces give much firmer forms than jigsaw-cut curves. They well suit its softer, more approachable surface. Curves can look silly made up out of planks, but when they are curves of firmness, of structural meaning (as a wooden boat is) then they look really wonderful. I am not averse to curved wood, but the curves need strength. If you steam or cut curves by hand, the limitations of tools and materials give this strength. If you have the freedom of a jig- or band-saw, they only will work if you have first learnt with hand tools.

Wood allows longer horizontal runs of windows without any visual loss of structural strength. Sometimes, even, the windows *are* the structure. Wood is for life above the ground. It needs a masonry base to root it in the earth — a heavy inward-leaning base, preferably part-covered with vegetation. The linear characteristics of wood can be exag-

Timber allows long horizontal openings to appear quite natural; not so masonry where vertical openings make better structural sense.

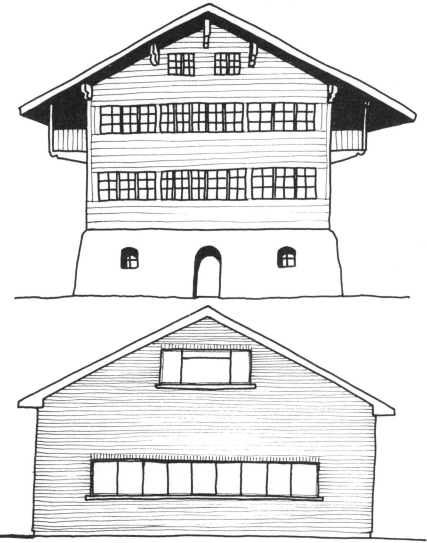

gerated or softened by colour — white fascias and corner boards emphasize the lines which enclose shapes more than do any other colour (except perhaps yellow or orange). Low pigment stains and, particularly, unstained natural weathering soften the effects of shape. Even very square buildings blend gently into the landscape when they are weathered grey: it is such a life-filled grey, quite unlike grey paint! Unfortunately though, it is not always the best thing for the wood.

I don't think that I am alone in feeling at home with natural materials. By 'natural' I mean of course modified nature. The tree is sawn and planed, earth baked into bricks and tiles and so on, but there is still a strong link between finished appearance (and sometimes feel and smell) and natural origins.

Natural materials are 'natural' for human environment. They help to give us roots. The need for roots has led to revivals of past styles of architecture — but, however skilfully they are recreated, when revivalist forms are built in modern materials — reinforced concrete, glass-reinforced plastic, imitation stone, wood laminates — they look as fake and hollow as they sound when you tap them.

One aspect of traditional building materials is that they are all bound by the scale of the human body: bricks are sized to be laid by hand, prefabricated panels by crane. Compared to ordinary concrete paving slabs (not my favourite material), concrete pavements cast *in situ*, sectioned only by expansion joints, are a huge step away from human scale. A large, simple roof can be at least acceptable if not attractive in subtly variegated tiling but is dominating and place-sterilizing if in uniform asphalt.

Anthropometric measurements such as the imperial system, and even more so the ell,[2] imprint the measurements of the body into a building. Our main concern however is how many body heights something is, how much above eye level, how many paces away, how much within or beyond our reach.

When we design things on paper we tend to consider dimensions arithmetically. 2.2 metres is a mere 10 per cent longer than 2 metres. In life, however, we experience dimension anthropometrically. A standard door opening is 2 metres high. 10 per cent higher it is almost at (common) ceiling level; we would hardly experience passing under it. 10 per cent lower and — at least psychologically — we need to duck under it. These few centimetres hardly noticeable on the drawing, make all the difference. (We can achieve both safety and threshold experience if we arch the opening so that it is high enough to pass through but feels lower.) Similarly, an inch more or less on the rise of a step makes a dramatic difference to the experience of going up or down stairs.

Small measurements in relation to eye level are critical to views and privacy. A few inches in the height of walls profoundly alters our spatial experience. We also experience

[2] Ell: fingertip to elbow measurement. It is particularly useful when laying stonework as a quick guide to the size of stones needed to complete a course.

objects anthropometrically. We can experience a sugar cube within the hand, and something larger within our arm encompass, but when an object is just a bit bigger so that we can no longer see or feel it without walking round it, we start to experience it differently. Even the smallest buildings like bus-shelters are in this scale, but when designing it is very hard not to draw, model, experience and think of them as immediately comprehensible objects.

How big a building appears in the landscape is affected both by the proportion of roof to wall and by

Buildings of comparable volume can have markedly different perceived sizes: walls confront the observer and imply used space within much more than do roofs. Furthermore, perspective effects can reduce the apparent height of roof ridges.

the time of year. Walls confront one whereas the roof slides away and also has a perspective reduction. The gesture of a steep roof can tie a building down to the ground whereas a shallow one with deep eaves can frame and emphasize a wall.

In towns, where views are often so hemmed in that we are not aware of the upper parts of buildings unless we look *along* streets, other factors are involved in perceived scale. Horizontal distance in relationship to event, textural scale and comparative sizes of distinct building units, vegetation and visible sky are more significant. Where we can see them, parapet skylines tend to increase apparent size and therefore 'urban-ness'; visible (therefore fairly steeply-pitched) roofs do the reverse. Where we may wish to

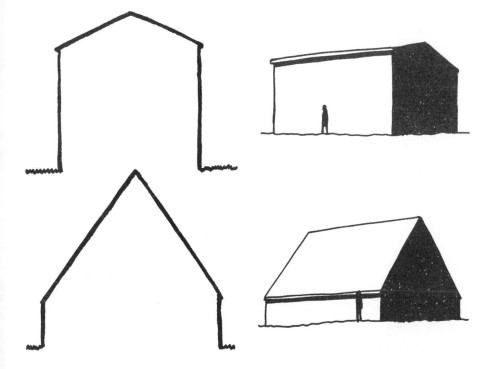

117

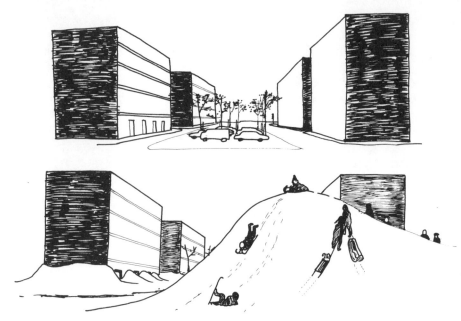

Snow banks can make a striking difference to the apparent scale of things.

reduce *apparent* density, with its close association with crowding stress, we may choose pitched roofs; where city-centre stimulation is a priority, they might not be so appropriate.

Seasonal growth or snow banks can make a striking difference to the apparent scale of things. The walls of traditional buildings with their low doorways and eaves were lower than some annual plants. In some three weeks of early summer or one night of snowfall, such buildings could change from focal points to the barely visible. Human life also, in its relation to the forces of nature, experienced the same dramatic rhythmic swings. Nowadays we have evened out these experiences; enlarging buildings, raising eaves, cultivating low gardens — often only mown grass — and imposing regular patterns of work regardless of season.

Nowadays many people seek to find roots in tradition, in tracing their family histories. The life-renewing rhythms of nature root us in time and place. But how many urban children even know that grass can flower? Every half month has a definably different quality to the preceding and following half month. Almost every week of the year is distinct, yet in many places you can only experience seasons. When I lived in London the months had no individuality — they were just summer and winter.

It is the progression of nature's rhythms in one place that is so rooting, centring, stabilizing. Travelling to find seasons — especially out-of-season, such as winter holidays on the beach in Tunisia or early summer skiing in Northern Scandinavia — is like buying vegetables out of season — and as crazily driven by economic reasons. To make money, farmers and market gardeners try to produce food *out* of season when the price is higher; often by the time the food reaches the kitchen it is hardly recognizable as anything that ever grew in the land — and neither is the tourist hotel in a fishing village or the ski-resort on summer-grazing pasture. To redress this de-rooting of everyday life, I have been asked to design

farms where city children can come to experience where food comes from, what happens at what time of year, how it is done, how *they* can do it — to find roots in life.

Places give roots to people, anchors which we need so much in rootless times when one after another codes of behaviour, established institutions, ways of looking at the world are called into question. Personal identity, marriage stability, expectations of employment — all seem so much less certain than they did to our parents.

Buildings threaten and destroy or add to and create places. Their first responsibility must be to add to places, to nurture the spirit of place — which in turn nurtures us. The interiors of buildings also create

The apparent size of a building appears to vary with the seasons. Annual plants can grow to human height in barely a month; leaf transforms a branched stick to a heavily laden tree able to completely conceal a scale-dominating object which appears large in the bare winter. Snow can also change the scale and focus of things.

inner places. Each room has a spirit. It starts with the architecture and develops through usage.

In the dark we can go into two rooms in a strange house — one is a bathroom, one a bedroom. We know instantly which is which; we can hear the acoustic difference. The architectural differences start with the senses. But there can also be rooms with similar spatial characteristics — say two identical prefabricated buildings in an army camp: one is a chapel, one a lecture room — a place for peace and a place for war-instruction. A difference in spirit begins to be noticeable. When the building has been used for generations — a church or torture chamber for instance — the feeling of this spirit is stronger. The place becomes imprinted by a spirit. However much it becomes a chrome and plastic city, who can visit Hiroshima without remembering?

As places build up their soul atmosphere to support a spirit of place, so too do rooms within one building: this room, for instance, or even part of room, is a hearth, the warm social heart of a *home*, not just the centre of a house.

Nowadays space is expensive to build. We design therefore in time and space; some rooms are multi-used. We think of time-space management. Indeed, sometimes I tell my clients that what they need is not more space but a different timetable. Most built spaces are empty more than they are in use, but we need to think first about the spirit of these places before we make any decisions about multi-use.

Some of these 'spirits of place' are resilient, allowing places to be used for many purposes. Others are more fragile. A cross-country run does not do so much harm to woodland and farmland but to a wild, empty, lonely mountain it leaves a long echo of *use*; not appreciation, but exploitation! Even amongst people who will not admit to anything spiritual in our surroundings, many recognize that the gambling machines cannot be satisfactorily moved into a meditation centre when it is not in use. In the same way, the protective tranquillity of a kindergarten is threatened when it is used for excitable debates about economic survival in the evenings.

Architecturally, what can we do to help nurture this spirit of place? Externally it is a matter of conversation between what already *is* and what we bring afresh with a *new idea* — an idea inspired out of the future, inspired from beyond the physical earth. Internally the occupants will be bounded by fixed physical restraints — walls, floor, ceiling. We need to bring in something enlivening, changing, renewing, something with a cosmic rather than just a human-usage rhythm; and that, of course, is natural light.

We tend to think of architecture as substance, but this substance is just the lifeless mineral vessel. Light is the life-giving element and both in quality and quantity it is absolutely central to our wellbeing. While light affects all aspects of mind and body, its effects are most pronounced upon the feelings. Just as warmth is related to activity and will (as you notice if you try to work in an overheated room) so

If we work sensitively with light, texture and space, even if only simply, even the more mundane rooms can be ensouled, can be welcoming, supportive places without the need to personalize and enliven them by adding objects, decorations, possessions. If we do not work consciously with these soul qualities we can hope to provide no better than the everyday norm: architecture which encourages the tendency to acquisitive materialism because of the need to humanize places.

light is related to the feeling realm, so much so that we often describe light in terms which we describe our own moods, like 'gloomy' or 'gentle', 'harsh' or 'warm'.

Inadequate light has been linked with depression and suicide statistics; however, we must not just think of light as a matter of physical quantity but as a life-bearing principle. We can enhance this life by how we texture, shape and colour the substance that frames and receives the light; for we cannot see light itself, only its meeting with substance. Some quite attrac-

tive materials drink up light, leaving a place gloomy even with bright windows; misplaced or shallow-set windows lacking tonal transition, harsh geometry and gloss reflection all tend in the same direction.

Light gives life to a room. There can be too much — window walled classrooms used to be the fashion — or too little. Rooms without natural light — here I think of the movement to classrooms with no windows at all — can have very disturbing effects upon physical, mental and social health. Laboratory rats in these circumstances attack each other or damage themselves. Some observers notice similar behaviour in those windowless schools.

The amount of window area we need to achieve particular lighting levels varies according to geography, orientation, climate and surrounding vegetation and topography as well as to the design of rooms and placing of windows. The amount of light we need likewise varies according to where we are in the world. City dwellers need more, so do those who live in northern latitudes or under predominantly grey skies, while towards the Mediterranean slatted shutters are used to darken rooms, creating quiet, cool sanctuaries from the outdoor heat.

It is easy to tell when there is not enough light in a room, when the windows are too small, but harder when they are too big. In Holland there are often a lot of plants and trees right outside the windows; the windows are big but the rooms gently lit. Candle-light gives life to a dark room which, poorly lit by a weak electric bulb, would be depressingly gloomy, and this is also the case with a sunbeam reflected off white-washed walls. To give life to a room it is much more a matter of quality than quantity. The human spirit needs this life-filled light. The soul needs it. Even the body needs it for physical health.

Sky-light from different directions and sunlight at different times of day have different qualities which breathe into our states of being throughout each day. Quantitatively west light may be the same as east. In quality they are distinctly different. The light of the seasons awakens us physically in the summer. In the winter its withdrawal awakens us to more inner activity.

Religious buildings — temples, cathedrals, stone circles — were built to correspond with chosen points in these great cosmic rhythms. Even today, simply for reasons of delight to the soul, we orientate windows to catch the sun-

Sunlight through vegetation gives gentle colour and modulation to the light in a room.

rise early in the year and to be filled with water-reflected light at midsummer noon and direct, deeply penetrating sunbeams in the winter. We can work with reflection. It needs care, however; reflection from snow can warm a solar collector or lighten a dark room, but the light is cold and there can be dazzle and glare problems. Reflections from mirrors can cause problems of deceptive space. Some designers like to play with this, but being deceived does not strengthen a sense of roots. We can also use reflection from natural materials and from paint. As I have mentioned, the delicate breath of lazure colour gives more life to the light than heavy opaque paint which emphasizes that static impenetrability of surfaces.

Once we think of reflection, we

think of material and substance. The right materials make a building. In the days of black and white if I photographed an attractive village street the photograph would often show mediocre architecture. The colour, light effects of sunlight and materials, not to mention its unphotographable sensory richness, *made* the place.

Materials and light are two completely opposite poles which belong together. Thick walls with sunbeams through deep windows, dark rocks in luminously still water, trees fringed with light against the sun: these joy the heart. The unphotographable because they are alive. Light and matter is the greatest of architectural polarities — the polarity of cosmos and substance, one bringing enlivening, renewing rhythms, the other stable, enduring, rooted in place and time. This polarity is the foundation of health-giving architecture, for the oneness of stability, balance and renewal underlies health.

The ancient druids worked with this polarity with rock and sun, for in the tension between them health-giving life arises. I also try to work with it in a qualitative way, and it is sensitivity to qualities that has led me in this direction rather than thinking my way. I started just by having a *feeling* for these things. I have therefore made a lot of mistakes, but the process I have gone through is similar to that with which one needs to nurture a spirit of place.

It starts with developing a feeling for what is the appropriate mood, then building a strong soul of a place with materials and experiences of appropriate sensory qualities. It starts with the feelings; architecture built up out of adjectives — architecture for the soul.

10

Building as a Health-Giving Process

Many people regard building as just a thing that happens, something which has nothing to do with health. In any case, health of what? How can the normal building process be considered unhealthy?

Building work is predominantly one-sided. It involves little more than the intellect for the managerial (and all too often, the architectural) side and physical strength and manual skills on the workers' side. Not surprisingly buildings built like this are sterile.

The whole process is one of materialization of ideas, often too fast and too far: too fast because the idea often becomes concrete and inflexible before it has met and conversed with the requirements of the surroundings and people. The buildings that typically result are imposed on, and damaging to, the environment and social fabric. Too far because decisions become dominated by monetary considerations. So do relationships — indeed conventional relationships in the building industry are governed by the principle of gain. There is a ten-

dency therefore to try to get the best out of any situation, to get as much out of it as possible. In other words, relationships are exploitive. Nobody likes being exploited and it does no one any good.

If we wish to develop a health-giving process we must start with the recognition that a human being must be meaningful, whole and nourished. To be meaningful one in some way gives benefit to others — one gives; taking cannot make one meaningful! Giving is not the same as imposing; taking is not the same as receiving. One is a selfless outward gesture, the other egotistical. In land and townscape as in human society a nourishing gift gives meaning to a place or a person whereas exploitive taking denies it.

Yet neither what we give nor what we experience through our work can be healthy unless we can effectively involve our whole being. Healthy work engages mind and heart as well as the hands. Also for the ultimate users of our product this has repercussions.

The extent that builders understand and care about what they are building shows up in the performance, quality and feeling of the end product. All too often we see the negative effects. People say, 'you can't get good workmanship these days . . .' But why should we expect good workmanship when we offer nothing to inspire and nourish the worker? Volunteer and self-build building provide opportunities more or less denied under the contract system for work to engage mind and heart as well as hands. Elsewhere, the client can rarely afford an artistic input, and when he does this is provided by specialists. The contractor makes his profit by using tradesmen who know what to do so well that they don't need to think. Their feelings have nothing to do with the job.

To be whole the polarities of one's being are brought together — the intellect and physical actions are brought into a harmonious relationship by the realm of aesthetic and moral feelings. Indeed, balance and harmony are vital to health — in the individual as in society or ecological communities.

Harmony does not occur when polarities clash. Just as when unrelieved planes meet at right-angles without mediation, there can be no harmony when fully-formed inflexible ideas are imposed onto a landscape or a future image onto a townscape which is evolving from the past through the present.

In the same way that harshly-

The fixed idea imposed as object: the individuality of place, the users and the stream of time through past to future are of no concern to such a building.

meeting architectural planes or building and surroundings can be brought into conversation with each other, so can architectural intentions with the ideas, sensitivities and skills of the building workers. Both internally and externally, roof and window shapes need to speak to each other. Internally ceiling shaping, view focus, light, and all the rest of the room need to join in this conversation; externally a whole range of elements, shapes and spaces will be involved. Naturally I try to achieve this sort of thing on paper well before building starts, but I can't see everything on paper. I consistently find that the fine-tuning can only be done on the spot, at full scale, involving the people who are working there, using the building as its own constantly evolving model.

Every building situation is unique. The building's relationship to its surroundings is unique; its users, even if we do not personally know them, are individuals. If designers live up to their responsibilities they must listen to the unique requirements of each individual environment, each particular set of users. If not . . . we have seen enough mass housing repetitively imposed upon the land and townscape! Once one listens like this, it is quite clear that no two sites, no two users are the same. They may share similar characteristics, but they are not the same. No model can fit in different surroundings; only in one can it be appropriate. While we can look to models (historical examples or places we know such as our own homes) to inform us and broaden our experience, we cannot merely repeat them.

An important characteristic of models is that they can be adapted. We can see how something looks this way or another way. It is quicker and easier when the design model is small, but there is much we cannot experience. It is really a form, not a space — something we experience from outside, not from within — an object, not a volume within which to live. Only at full scale do so many things become apparent. There is no undue difficulty in modifying the design as one goes along. Within structural, constructional and legal limits it always feels to me the natural way to go about it: the problem is to find opportunities to be able to.

The historical development of building into the contract process has been a process by which a design has become frozen. Everything has to be described by the contract documents, and these are confined to the level of perception that we can achieve from paper or small models. In other words shape and form — not space.

In the conventional building process, time costs money. Flexibility takes time and also makes rigid pricing difficult; it also adds financial risks! So everything has to be fixed and put on paper — and, as a matter of course, we accept the disadvantages.

There are other forms of contract. A lot of small builders will only work on a time plus materials basis. If honest, it is fairer all round than a fixed price; the builder neither needs to take risks nor cover unknowns by the 'double-it' rule. (For

Every client, occupant, user, even those not yet born, is an individual, a human person, not a feelingless statistic to be packaged. They need their own places as houses for the soul, not as boxes for the body.

Time plus materials contracts depend upon trust, honesty and flexible budget limits. There are, however, no monetary incentives to tighten management efficiency, spend the time to shop around for materials or subcontract to more competitive specialists. On the other hand achievement-related bonuses put quality of work at risk.

Between these contractual extremes I have tried to develop a pricing system that gives room for qualitative work and where unknown costs or over-competitive pricing need not bring disaster to client, contractor or workers. What I call target pricing, is calculated on a time plus materials basis:

Estimated time × rate per hour = target price

If work takes more or less time than estimated, the rate can be decreased or increased up to a mutually agreed margin, although the price may not exceed the target if work is quicker than expected, not be less if slower. Undue profit or loss is therefore restricted to a level mutually agreed beforehand.

example, non-standard door: double the price hang and fit: double again. Whereas actual extra costs are 50 per cent extra labour on door but no more on materials and hang/fit time. Total extra cost to builder is 12½ per cent, but to client is 100 per cent.)

For example:

Estimate: 100 hours × £5/hour = target price £500
           margin: £1/hour

Actual job:

A.   140 hrs × (£5 − £1 =) £4/hr   = £560

B.   105 hrs × (£5 − £1 =) £4/hr   = £420 ⎫
                                       or £500 ⎬ = £500
                                                ⎭

C.   85 hrs × (£5 + 1 =) £6/hr   = £510 ⎫
                                       or £500 ⎬ = £500
                                                ⎭

D.   75 hrs × (£5 + 1 =) £6/hr   = £450

This system is only a start and is by no means immune from problems, such as dependence upon honesty or meaningful (and laborious) records. Contractual systems are not very exciting to get involved with but they do have a significant influence upon working relationships and the quality of the end product. There is a real need for innovative forms of contract to release creativity and work as an (albeit paid) gift from their present day straightjacket.

However necessary contractual working relations may be, all systems to some extent deter workers from involving themselves artistically. But there is another way. If time is not given a monetary cost, it can be used to allow the design to evolve on site, to develop potentials that only become apparent at full size — indeed in every way to improve quality, visible and invisible, in so many ways. If inspired by beauty, self-built buildings have this opportunity. I have a client who in his own words was 'building sculpture to live in'. If inspired by pecuniary motives, the opportunity no longer exists for in that world 'time is money'!

This opportunity is enhanced in voluntary projects as time spent is money saved, and time spent can be used artistically. A group of which I was part bought the building, derelict for 20 years, that is now Nant-y-Cwm Steiner School.[1] After purchase, we had only some £36 to finance major repairs and alterations without which the

building was totally unusable. We had only two options: start work on those jobs that were 95 per cent labour or give up the whole project.

We started work, initially two of us, and through working opened the door to donations and help from people we had never even met. The result was not only that the school came into being (which without the gift of work it could not have done) but also that financial stringency became transformed into artistic work.

In this project the flavour of the brief was established by the qualitative and economic requirements. As with the education the school would give, the building should provide an environment nourishing to the child. Children should not be forced into a mould by the architecture but have the possibility to move, live, imagine in their own child-world and to receive attention as individuals. The building should therefore show individual attention throughout, all woven together to create a harmonious, gentle environment for the child. This then requires individual attention from the builder. It is not possible to design for individuals without individual attention. Imposed standard details have no place in such a building.

But what does 'individual' mean? If made by a discerning hand, no two door handles will be the same. Similar, perhaps, but not the same. No two doors will be the same. Each is the result of the conversation between wall-shape and opening, between one space and another, meeting at a portal, a punctuation point to our movement and pro-

---

[1] Photographs show this school throughout the book — see List of Photographs for examples.

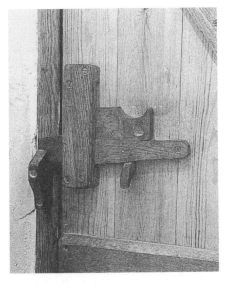

Once we recognize that every situation is unique and once builders work not as mechanical executors of other's orders but as artistic individuals even every door handle will be subtly different from each other.

gressive experiences, an open or closed eye — a door.

The economic requirements were more straightforward: build at minimum cost! This inevitably meant voluntary work. Labour therefore was free, allowing us to incorporate much labour-intensive handwork, to give every element its own individual attention, to take time to bring it into conversation (even song) with other elements. Therefore the economic requirements interpreted in this way supported the fulfilment of the aesthetic requirements.

Working in this manner has profound implications for the people involved in the process as well as for the building. It soon became obvious that gift work is sustained by inspired will. It is vital therefore that this inspiration is nourished or soon there will be no volunteers! While gift work is commonly seen as a one-way process of giving it actually required the work situation to give *to* the volunteer, and the organizers have the special responsibility to arrange the work to enable this to take place.

As foreman of a volunteer building project I have learnt to try to present even small jobs so that their part in the performance of the whole can be seen and so that workers are led into understanding why they are doing something in a particular way. I explain for instance *why* as well as *how* intricacies of damp-proof, thermal or acoustic detailing are so important, why some jobs must be done in a specific order and so on. It is important that each day is marked by visible achievement — this often requires organizing work on a teamwork basis so that everyone, regardless of skill or strength, can contribute and at the end of the day see clearly what has been achieved. Blocklaying for instance can absorb five people: to supply and cut blocks, mix mortar, lay it on the wall, place blocks, prepare and level, and check.

Perhaps most important, however, is the cultivation of aesthetic involvement — and I mean cultivation because the seeds already lie within each of us. Building is essentially an artistic process on a large scale with more parallel functions than painting a picture or modelling a sculpture. It is out of our aesthetic attunement that *how* to do something — plaster a wall for instance — emerges. A plaster surface can be dead, limp, without strength, or as life-filled and harmonious as the firm forms of landscape. I can describe how to do it in words but only the experience of actually *doing* it can make it live inside oneself and enliven the aesthetic sense.

This perhaps is the greatest gift to the volunteers, that they develop in themselves, in addition to manual skills and meaning in work, a sense that any work can be artistic.

The benefits for the building itself are no less significant. Fundamentally gift work is the reversal of the normal contractual approach. Normally inspiration — the idea, the client's and the architect's vision — becomes progressively more and more materially defined — drawings, specification, bill of quantities — until it is solely a monetary description — the contract. The contractor seeks a profit, the work-force a wage; the project is coloured by the principle of monetary gain, yet it started as inspiration. Spiritual values have become reduced to material values, qualities to objects, adjectives to nouns. In biblical terms, bread has become stone.

With gift work, it is the other way around. Building materials can, through gift, be raised to a work of art. I don't believe it is possible to create works of art without gift. One gives oneself. It is possible (or even sometimes necessary) to be paid, but this is always secondary.

Raising is the key element. If the building is to be nourishing to the soul, the material elements must be raised, by artistic means, to the spirit. We can look at food in the same way: to be nourishing, mineral elements are raised through the processes of life to become nutrient substances. If only half-raised, protein is poisonous, as is for instance spinach grown in warmth but inadequate light. They are then further raised in the kitchen to provide a truly nourishing food — half raised they are uneatable. Cooked without love and delight, they only fill the stomach, they do not *nourish* us. Nourishment from the environment is the same.

This approach imbues material substance with spiritual values — art, inspiration. And this has benefits, visible and invisible, for the building, those who work on it and its future users. The visible benefits are obvious in the qualitative, aesthetic sphere. The intangible, invisible quality of a building is also quite different if it has been built for profit or gift, without or with love.

To the practical 'realist' these

benefits are not materially measurable, but there are also economic benefits. The school I described was built at approximately 17 per cent of the estimated contract price; sometimes this 'impractical' approach is in fact the *only* practical course. Indeed in this case lack of money actually reinforced the artistic impulse. Not a complete absence of money, but certainly there was no surplus! Lack of money has had its negative side too — much needless drudgery such as mixing cement by hand, inefficient sequences of work and many delays — but it has not restricted the real *wealth* of the project, the extent to which artistic, spiritual values can raise matter.

When people say, 'there isn't the money to make a building aesthetically satisfying', it is just not true. What they mean is that there is not the will or priority. Money for them is the first priority, aesthetics follow — and we all know what happens!

It is not that supportive money is unimportant — far from it. However, when it becomes a ruling consideration, when the project is for the accumulation of money, practically the only purpose (because it rapidly becomes the dominant one) is the pecuniary one. Users and environmental responsibilities become

Buildings cost time. If you build your own house with all its components it may take two or three years. If you buy it built by others the price is comparable in the time taken to earn the money. We take it for granted that people built beautifully in the past, something we cannot afford today. How could they afford it?

Perhaps half of the cost of a building goes on services and fittings, something our ancestors never thought of. Many of these comforts have been bought at the price of aesthetic quality, for today we cannot afford to do without the comforts but we cannot afford the aesthetics — or so it is often claimed.

secondary — and, insidiously or overtly, the results are inevitably destructive.

The approach to work as gift not gain is less easy, though always possible, with paid work. The concept of measured exchange, of buying, tends to enter in. 'I do so much for so much pay' — 'you are expected to do this because I am paying you . . .'

In every sphere of society, in every sort of work, it is the *approach* that is crucial: art or profit, service or exploitation, need fulfilment or market opportunity, material or spiritual value. Just as the gift principle can be applied to all work, so can the ideal of all work as an artistic — a sacramental — act. But most work is not outwardly encouraging to such an attitude; the individual needs to have travelled this inner road himself. Not all of us have.

Gift-work projects provide a more supportive framework to this inner growth process. The benefit charities bring to society is not limited to the services they provide; it also gives to the individuals who participate, the opportunity to experience work as meaningful gift and creation as service rather than indulgence — triggers for the reappraisal of values. Anyone in any situation can go through these growth processes, but without support far fewer will do so.

The opportunity for aesthetic performance and achievement involvement, is a cornerstone of this supportive framework. Projects need to be managed differently to maximize this aspect. The emphasis needs to be more on the aesthetic consequences of work than on high productivity. Since work is given, this is a necessity as it supports inspiration, the essential key upon which voluntary projects depend. There are, therefore, very practical reasons why wholeness must replace atomism in management thinking and working relationships.

How does this work out in practice? Converting old buildings gives plenty of scope for mocking-up, chalking out, listening to sound transmission and so on, as I have described; we can *see* what needs to be done next. If we listen to the needs of the existing building meeting its future, work can be the artistic response to the situation. New buildings are at first sight more constraining. The integration of the design tends to appear complete and fixed on paper; but the paper design need only be the starting point.

On the Steiner kindergarten I am now building I found that the exact plan shapes, the curves of the walls, the size of play alcoves, could best be established at the stage of laying the first bricks. Only at this stage did lines on paper become boundaries to space. Only by daily observation of sunlight does it become clear which trees should be felled and which retained. Only by standing at future window openings is it clear exactly where they should be located for sunlight and focus of view. Just where and how the different geometrics of the roof meet is much clearer on site than on the model.

Allowed to evolve on site, the building now starts to take over. If it didn't, we could build nothing better than a good plan. What work of art of any worth is an exact,

larger scale model of smaller sketches — how can it be art if it is merely a fixed idea? The painter's sketch, the sculptor's maquette, the architect's drawings are only a way in. Then the building starts to grow as a real living being — with its own suggestions and demands.

There are a lot of things which can only be brought to fulfilment by design on site and this means that builders *must* be brought into the design process. In fact I consider it vital that they are involved artistically. Indifferent architecture built with care and artistic involvement can become a beautiful, soul-nourishing environment. Excellent design built without care or concern *never* can be.

In any case I take the view that builders often know a better way to make something than does the architect. One often has to be quite careful in this. It is only a step to go from how to *make* something to how to make it *look* as we want it.

What can be merely structural problems on paper can on site be seen to be opportunities for artistic development. Here the builder designed the details.

133

Together we can discuss, gesture, draw, mock-up the *exact* curve, the exact view from a window, the sort of quality of a door latch, of plaster texture and so on.

There is no better way to prepare oneself for these conversations than making things oneself. Here one learns how the *material* and the *act of making* suggest the way that the appearance, feel and so on should go. Also when the builder says, 'It can't be done!' one is in a better position to say, 'Yes it can: I have done it — and I did it like this!' After all, all builders know that all architects know so little about how to make things that they design

Architects tend to think large. Unless details are standardized for repetition or individually described — involving an uneconomically huge mass of drawings — it is up to the building craftsperson to make the little bits. But it is just these little bits that are the contact points between users and buildings. For small children especially, they give the opportunity to 'use' the building in their imaginative world.

things that can't be built! It should not be a case of passing down orders but of *agreeing* things together. If we disagree I can give an order and the builder will do what he wants to anyway as soon as I go away!

This isn't the conventional way of doing things. Conventionally the architect's great ideas are executed by others and in the process compromised by constructional requirements, by builders, by users and eventually by ageing and weathering. The building can be completely ruined by this process. I take the unconventional view. I feel that the life-history of a building from clients' need to a well-used old age should be the history of *continual improvement*. It should get better and better at every stage.

Constructional design brings our attention to the details people will actually touch, sit on, use, bump into. Building construction allows us to see what we could not visualize on paper and to develop its potential. Building craftsmanship brings care into a building and gives it a soul. Users bring life and spirit into a building. Time brings maturity, richness and increased harmony with surroundings. But none of these are isolated compartments: each stage brings something new to add to what is already there, a conversation between what has become and what is becoming.

Design is the process by which needs are worked out as practical solutions, and it depends upon this conversational principle to maintain its balance otherwise it becomes a compartmentalized activity in itself and its results have no relationship to place or people, to builders, users or initiators. I certainly have gone out of this balance far too often and done things on my own, neglecting conversation when it was most needed. To be balanced this process needs both overviewers and localised involvement. Neither overall planners, specialist consultants, builders nor users can be left out.

If we think of the process as a whole we can see it as a cycle through spirit and matter. It starts as an abstract idea. This becomes more and more condensed into substance — a building — and this in turn becomes more and more filled with life, by its builders, then by its users.

At least, that is how it should be. What we often see, however, is a progressive descent of that which started as inspiration into more and more lifeless material form: the architect's inspiration ossified into contract documents, quantified in monetary terms and eventually built as an exchange of building substance for monetary reward. But it need not be like this; an interweaving upward stream can also flow. On the one hand the inspiration finds its feet on earth, on the other matter is imbued with spiritual values by being raised artistically. Both matter and spirit need each other — neither is whole without the other. But this can only happen when the work of *building* is approached artistically.

It isn't usual for builders to bring artistic values to their work, but it *is* possible. In my experience, however, it is *only* possible if they are involved in the artistic process. Certainly, the less builders are involved, the less they can be expected to care. Yet it is the artistic sensitivity of the making hand that is *vital* to the process of ensouling buildings — at least as vital as the architect's skill. Just as it is the architect's task to bring soul qualities into the rational world of regional planning, it is the builder's hand that gives life to the architect's plans.

An important effect of letting the design evolve as potentials become apparent and are developed is that the building process and the building itself develop a kind of life. Hands which work with loving feeling imprint a kind of soul into the building. You can go into the empty unfurnished building and already feel 'it has a soul'. Go into an unoccupied machine-made building in which the workers had no aesthetic involvement: it waits to be given soul by its future occupants, and they will plant their qualities upon no foundation — it will not be a conversation, part of time's continuum. It will never be as much a 'home' as if the building started out 'ensouled'.

This is not to say that some people don't interpret this process as freedom to depart from the design, or, deaf to the incarnating theme, impose their own personal preferences. In every building that I revisit I am acutely aware of those things which were not done as I would have liked them. But it is only *I* who notice them. What others see, appreciate, and misdirectedly give me credit for, results from *how* the building has been built, the work in which others have *gone beyond* what my design only started off.

Even before occupants breathe life into a building, even before it is finished, the process of ensouling can be well advanced. Had the (contracted) building workers not been aesthetically involved in their work, I doubt that it could have been so.

Any credit for *how* a building has been built is due to the building workers. Yet how many get a mention on architectural prizes? Recognized or not, however, what this process means for them is that more even than completing the physical building (the noun) they are building the qualities (the adjectives). As best they can, they are building something beautiful, nourishing. Just as, for instance, to poison others with words poisons oneself, so to make for others something nourishing, brings unlooked-for nourishment to the building workers. Good nourishment is essential for good health. But no product will nourish either makers or users unless it has been made out of the spirit of artistic gift. To find ways to make this spirit accessible in the daily working situation is a task vital to the health of society.

We have come to take for granted that buildings are provided ready-finished by others. The less we are able to do ourselves, the more dependent we become. Dependency is a step towards social malaise. The wealthy can buy their way out, commissioning others to design and build their homes and workplaces. So can the skilled who can do their own building. Self-builders are estimated to build more houses than any firm in Britain, though very few start low on the ladder of privilege. My experience with volunteer building, however, demonstrates that even with no previous skill and with low expense it is possible to break the links of individual-suppressing dependence, to afford and achieve surroundings that nurture the soul, to build self-confidence, community and, almost coincidentally, learn employable skills by the back door.

11

Healing Silence: the Architecture of Peace

Healing is a process that can only take place from within ourselves, but this process can be triggered and supported by things and actions outside us. We can, therefore, talk about healing environments and healing qualities of environment. Of all the healing forces in the God-given world around us, silence is perhaps the greatest. We have seen the health-giving effects of processes, activities and material qualities, but silence is neither process, activity nor object. It is. . . silence.

But what is silence? Is it complete absence of sound? Where can we go in the world and find absence of sound — no wind in grass, no distant clink of rock, no lap of water? Sound means life; in quiet places, the ears sharpen to listen for it. We even start to hear the sounds of our own body.

There is a lot of difference between a resting and a dead body. A dead animal looks different in the landscape to one lying down; the wind plays with the hair as a lifeless surface of something immova-ble. This is the silence of death. To experience literal silence you have to go into a special sound-absorbing chamber — it is a strange feeling. Sensory deprivation experiments have shown that if all the senses are denied stimulus, the life processes are brought into a crisis so acute that within seconds a risk to life develops. Literal silence is not life supporting: it is the opposite.

Or is silence the absence of noise? Even noise is hard to define: is it insects on a quiet summer's day, waves on rocks, wind in trees or over snow? But there is plenty more noise than that around us. The average house is full of noise-producing equipment — refrigerators, deep freezers, central-heating furnaces and pumps, ticking electric clocks and so on, all dead mechanical sounds, not sounds of life like speech, music, crackling fire, wind in the chimney, rain on glass. Out of doors cities have constant background noise you cannot get away from. In the countryside how far do we have to go not to hear a hi-fi, car, chainsaw, milking-machine or aeroplane? When you listen, almost everywhere within easy access of where we live there is mechanical noise most of the time. In this century silence — freedom from mechanical noise — has become a threatened species, extinct in many areas.

We live in a noisy world: whether we notice it or not, noise affects us. Physiological effects, starting at 65 dBA with mental and bodily fatigue, are well established.[1] This is typical city noise level.[2] Main-road kerbside noise, typically at 75 dBA, is over twice as loud and motorways nearly double again at 83 dBa.[3] Street noise can reach 90 dBa causing heart stress.[4] Much lower levels, such as background fan noise, interfere with sleep, digestion and thought.[5] We easily become conditioned to low-level noise and don't notice it at all. That it causes tension is however demonstrated by the great relief we feel whenever it suddenly stops.

Noise, in other words, is harmful to human health; it is a recognized environmental pollutant. There are well-established techniques for noise-abating design. Distance, obstruction (for instance by walls, banks, buildings), absorp-

tion (for instance by vegetation, which can also act as a fume filter), zoning of sensitive and tolerant areas and masking (for instance by rustling leaf, moving water or living sounds) can all mitigate outdoor noise. Where aggressive movement such as fast traffic is the source of noise, it often helps to screen it visually. Intermittent noises such as trains on the other hand are less of a shock if you can see and hear them approaching. Noises from living sources such as school playgrounds can be less irritating if you can see what is going on. Outdoor noise penetrates indoors mainly through openable windows. When noise and air pollution sources coincide, as they often do, windows facing this way can be sealed (and double-glazed, absorbent lined, etc.). Indoor noise can be reduced with absorbent materials, room shape and control of noise-making at source.

There is of course more to noise control than I have here outlined, but however thorough our measures we cannot hope to achieve silence. With, for instance, triple glazing and absorbent indoor surfaces we can make acoustically dead environments, but that is not the same as silence. Yet silence is something we need to have access to, for while noise is stressful silence is healing.

Where in the world can we go to find this sort of silence? And for those who can afford the expense, how much noise does travelling there cause? One place to go is within oneself. Many seek inner silence through meditation, but it is not easy to keep inner noise at

[1] SV Szockolay, *Man-environment sonic relation*, (Course notes: E 13) Polytechnic of Central London, p.9.
[2] Typical values, 10 per cent of the time 7am-7pm all use zones in Inner London — *Traffic noise: Urban Design Bulletin 1*, GLC, 1970.
[3] Every increase of 10 dBA represents a doubling of apparent loudness.
[4] Ian McHarg, *Design with Nature,* Doubleday/Natural History Press, New York, 1971, p.195.
[5] David Wyon, *Det Sunda Huset*, p.196.

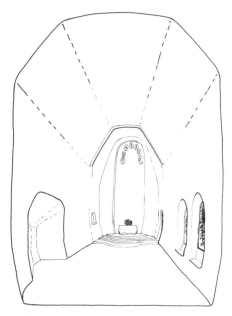

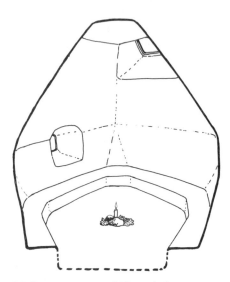

Chapel — different elements on either side of the axis balance each other.

Meditation room — different light sources to enliven a central almost symmetrical space.

bay. But if we are to design healing environments we need to create qualities of holy silence that are accessible for all, not just for globetrotters and meditators but especially for those who lack the outer or inner means.

Even if we can't *define* silence, we can recognize it. Gentle, unobstructive, calming, life-supporting, holy sounds allow us to be quiet within: eternal sounds, sounds of the breath of air, the quiet endlessness of water — definitely not sounds of the ephemeral moment however calming. Cows chewing the cud and bumble-bees droning are calming, almost soporific, but they are not eternal. Silence, tranquillity and the eternal have a lot to do with each other.

It's even harder to define silent architecture but likewise easy to recognize it. There is dead silent or living silent architecture. To create living silent architecture we need to understand and work with the essential qualities of living silence: the gentle, the unobstrusive, the tranquil, the eternal, the life-supporting, the holy.

As a foundation of tranquillity we need balance. This often means focus and axis. Symmetry is rigid, rigidity excludes life. Balance is life-filled and breathes from one side to the other. Balance is also a matter of scale and proportion. Rooms can be quite small — monks' cells were often little more than the space to lie and stand in. The smaller a room is, the more modest, plain, ascetic and quiet it can be — furniture is an intrusion. Such a room is for a specific purpose, but not a silent place *within* the stream of daily life. If the proportions, textures, light and other qualities are not just right, a small room is a trap, a larger one can often get away with it although you can start to rattle around, and its silence can begin to feel empty. Too large a space can be too awe-inspiring. The human being is too insignificant beside the power of architectural scale. Those cathedrals that are places of silence (and there are not so many, for more are places of awe) are not the largest ones, their scale reduced by the way they are built of tiers of elements. The gestures of the romanesque ones tie them down firmly to the earth. Imagine such a cathedral plastered and painted uniformly — in simplicity its size would be too strong — certainly it would not be silent!

Proportion determines whether places can be at rest or whether they have a directional dynamic and the feeling that goes with it. Awe, expectation or soothing can be produced with upward, forward or all-round horizontal emphasis. Proportions at balance reflect balance in the human body and induce a mood of balance in the soul.

Proportions that are too high, too wide, too long — like lines that are too dynamic or spaces that are too strongly focal — risk being too compelling. I want to leave the occupant free. I try to be careful, therefore, not to have too strong an emphasis. Indeed for a place of silence I try to underplay the architecture generally so that it is not intrusive. This means a certain simplicity. Simplicity, enshrined in the modern movement, is often experienced by non-architects as boring. Some buildings need to be less simple, some more so. Places of silence need to be simple — but

how can reverent simplicity be achieved without boredom?

I approach simplicity like this: the space can generally be entered and focused axially but slight variation from one side to another, slight ambiguities in form and, most particularly, living lines (flare at the base of the walls, curved qualities in the almost straight and straight in the curved and so on) give the space a quality of life — so too with straighter, but not colliding, lines does the texture of wood, even if its colour variation is muted by stains or lazure veils. This life is further enhanced by the light. Where the windows are placed, how they are shaped, how the light is quieted — for instance divided by glazing bars reflected off splayed windows or filtered through vegetation — can enhance the interplay between daylight, sunlight and reflected light which is so crucial to the mood of a room.

Where windows can be deep set into the wall, the light reflected off the reveals not only adds to room illumination, but intercedes between the brightness outside and the shade within, giving a calmness to the light. A frameless window and a softened but firm and balanced shape and soft plaster texture can add to the mood of calm silence.

Light needs texture to play on. Again I am looking for a life-filled, but unobstrusive, gentle texture. I commonly use hand-finished render.[6] This can bring gentleness, life, conversational softening of changes in plane and — because of the absolute necessity that the plasterer is aesthetically involved — soul is impregnated into the room. This certainly is not the only material, nor is it everywhere appropriate, but where it is it is one of my favourites. Being applied to block-work and requiring more sensitivity than skill, it has the additional advantage of being cheap and well-suited to gift work or self-build.

I've been in spatially simple rooms which lack any life in their texture. Smooth-plastered, smooth-painted rooms, even the woodwork gloss painted. To be alone, quiet, in such a room is to be in a prison. You *need* a radio, hi-fi or television for company to fill the empty space, to bring a kind of life. I aim to make rooms in which you do not need these supports, rooms that will be alive with sunlight or candlelight, birdsong outside or with grey dawn, twilight and silence.

Texture-less rooms need wallpaper or colour schemes to give interest, to paint a superficial individuality upon their surface. I use colour for a different function, so different that when a client says,

'I have these curtains, I want a colour scheme to go with them', I am at a loss to know what to do: I use colour to create a mood. Yet often, for silence, the indoor colour I use is white. White, justifiably, has a bad reputation; it's the colour people use when they can't think of anything else. But it is the colour I use when I don't *want* anything else, when I want silence. Some people think white is not a colour, but the right white (there are many — think of the difference between lime wash, emulsion and gloss paint, not to mention all the different colours of white) can sing! White is the mother of all the colours — it has in it all the moods that each individual can develop as an individuality — white can be calm, life-filled, joyous, timeless, whereas blue can only be calm and risks being cold or melancholic; orange can be full of life, welcoming, but risks being too forceful, even discordant; yellow can be joyous but risks being too active; I have never seen more eternal qualities than in Vermeer's paintings, yet brown risks being too heavy, dark, oppressingly entrapping.

However, where the room or window shape is rectilinear with smooth surfaces and sharp arises, I would certainly not use white — it would be altogether too hard. In such a room white would emphasize any noise. Research on colour and perceived noise does indeed show white rooms to sound loudest.[7] When we use it we

[6] 9 coarse sand: 2 lime: 1 cement, applied not by float but with a round-nosed trowel so as to obtain gentle undulations without tool scoremarks, finished with a (gloved) hand when it has started to firm upon the wall (about an hour later but depends on conditions).

[7] Kenneth Bayes, *The Therapeutic Effect of Environment on Emotionally Disturbed and Mentally Subnormal Children*, Gresham Press, 1970, p.33.

therefore need to be particularly attentive to qualities of shape, texture and light or indeed it will seem noisy. The quietest colour for a room has been found to be purple. In ancient times purple was not a colour anyone could use, its use on clothing restricted to a certain spiritual rank. Even today, less sensitive to the 'beings of colour' as we are, it doesn't seem quite appropriate for everyday use; a purple kitchen doesn't feel quite right.

I use particular colours when I wish to emphasize a particular mood. Red can bring warmth, stimulation, passion and aggressiveness.[8] With all colours, associative qualities such as coolness with blue are bound up with physiological effects upon the organs and metabolism. Yellow for instance can bring light to a sunless room; it can also bring vitality and cheerfulness. Green is calming and refreshing; it is the colour of surgical gowns and actor's 'greenrooms'. The meditation room (shown earlier) is to be lazured in bluish purple.

In therapy, coloured light has been shown to be more effective than pigments.[9] Coloured windows only feel at home in specific sorts of place (such as churches) but coloured light can be achieved by reflection. Opaque colours are forceful and dominating, translucent lazure is therefore both more acceptable and more effective than opaque pigment. Brick tile and timber with dark rich weavings bring warmth. We even designed a theatre to be painted inside in grey (not a flat grey but one made up of thin veils of red, blue, green); this was also to be a focal, unobstrusive space, but not a place of silence!

Similarly I try to make silent, sacred rooms *plain*. These rooms need to be somehow above any more specific mood. When the circular meditation chapel (see page 73) was nearing completion it looked so attractive inside with its exposed radial rafters that many people wanted them left like that. I felt that they created a warm cosy atmosphere with noticeable architecture, appropriate for a living room perhaps, but not for a chapel, especially not for the silent, spirit-renewing focus of a retreat centre. I offered to pull the ceiling off if people didn't like it: fortunately I didn't have to.

In the same way that colours can be too individual so can material. The difference is that certain materials are the right materials for the place and it doesn't feel right — or may not be practical — to use other ones. In some countries brick, stone or timber is the only suitable choice and I have experienced wonderfully sacred places in diverse materials. Generally there needs to be very few different materials. Often I use only three: walls and ceiling of the same finish, running without break into each other and unified by a single uninterrupted colour; wooden doors and windows, unpainted but possibly stained or translucently lazured, and a texturally inviting floor of a colour to warm reflected light — usually wood, brick, tile or carpet. Unity of materials and colours has a quietening influence and for that

[8] *Ibid.*, p. 32.
[9] *Ibid.*, p. 32.

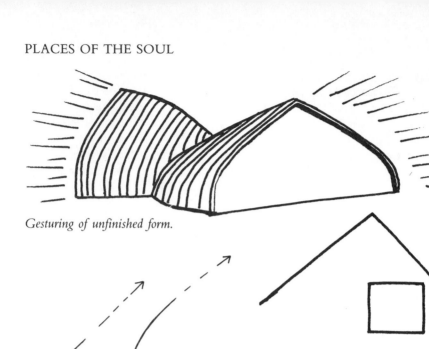

Gesturing of unfinished form.

In any line curved or straight, there is a dynamic.

So also there is in the shape and tension between two lines. Responding to one crisis can often produce another! To peacefully resolve these unfinished movements can take hours of effort!

reason they need to have sufficient life in them or the whole place will slide into lifelessness.

The shapes, forms and spaces need therefore to have gentle movement. The static resolution of the right angle lacks life. Dramatic or dynamic forms and gestures have an excess of force. To have both movement and stability, gesture needs to answer gesture in a life-filled, harmonic conversation — not repetition but resolution, transforming what the other says so that it is just right for its particular location, neighbours, material and function. Quiet harmony is the product of a quietly singing conversation.

Perhaps the most essential quality is timelessness. A painting can be timeless, so can a building. Obviously the painting has to avoid

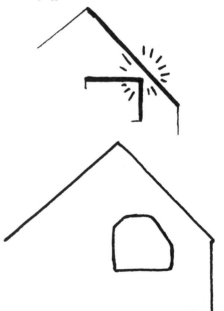

anything that finds its resolution in time outside the moment — like someone kicking a ball. The same with a building. This doesn't only mean qualities which are both traditional and modern at the same time; it also means resolution of the sculptural forces — of gesture, of gravity, of structural and visual tension. Dead things are stable, immovable, but they are left behind by time. The eternal lives in every moment.

It can help if one practises timelessness exercises. I like to paint uneventful balanced landscapes (of the soul imagination — not real ones) bathed with peaceful light,

Even a quietly enclosing curve still has a visual force — reflecting its structural force. The meeting of the moving against the rigid can appear unstable unless resolved by moderating neighbouring elements, angles and meetings.

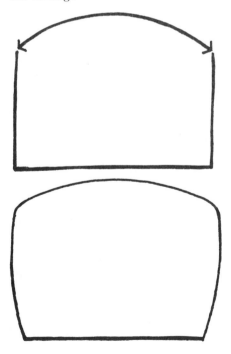

trying always to find that which is eternal, not momentary. I have mentioned the principle of balance, giving stability and permanence without rigidity. Buildings which belong in a place, which are rooted in the earth, can be developed to be timeless, eternal. Buildings which don't never can be. In addition to the shaping of walls and ground, as discussed earlier, planting at the building-ground junction and climbing plants on walls can help.

As far back as I can remember I have looked closely at how rocks rise out of the earth. Some are half-buried boulders, some are the protruding bones of the earth. Some mountains are the earth itself pushed through or draped with covering but now at repose. It is where they are at the firmest rooted and least dramatic that they are the most eternal. This as much as anything — the landscape I have grown up with — has sharpened my feeling for timelessness.

Other people in other surroundings may not be so lucky. For occupants therefore this means creating these qualities in the buildings we design. For designers it means developing sensitivities through exercise, observation and focused concentration. To be timeless something needs to feel inevitable, right — so much so that we can no longer imagine something other than the way it is.

The building for instance needs to be in the inevitable place on the site. That is not always easy! Sometimes a site asks for something somewhere, sometimes it doesn't. The hardest site I have ever had was flat, featureless, with only short-

I try to make buildings which belong in the place they are, which are rooted in the earth, which give one the feeling that they always have been and always will be. Places which have this eternal feeling can convey stable yet life-filled tranquillity in a way that those bound to a transient moment of style never can.

lived caravans on it; nothing to grow from, nothing to create a place between, nothing to relate to. Usually, however, listening to the place will give one a progressively strengthening conviction that this building should be *here*.

Buildings are not just in isolation. As we design them, we are also concerned with the whole entry experience. Externally we can develop this experience progression to enhance the inevitability of the building we eventually reach. In-doors we can carry this preparatory experience further until we reach the place to stop — a sanctuary of rest. We can enhance the experience by making physical thresholds wherever there is a change of mood by using darker, lower, narrower

Our ancestors knew well that the places we pass through affect our inner state of being. Typically a church was not entered directly from the busy street but after a series of threshold experiences to support the necessary inner preparation.

passages, cloisters, tree-overhung paths, leading to a portal — a substantial door with a heavy latch, which one is *conscious* of opening. Then, with a conscious step, one passes into another place, a place to stop — the place of calm, protected, enclosing. A glass box is not a place to find inner calm in silence. Its function is to wash one inwardly clean with the forces of the landscape. In more densely settled surroundings one can feel a bit as though in a display cage, certainly not at peace!

This progression of experience is made up of the same vocabulary that is available in most homes or places of work: thresholds, emphasized by portals, doors and latches, places to move in and places to stop. It can be enhanced by making these more conscious. I like hand-made wooden latches that you really feel and with a movement that gives you a conscious bodily experience of opening or closing a door. I like low (or broad so as to have a lower proportion) doorways with arched or shaped heads, low, dark, arched or shaped ceiling passageways, slightly twisting, leading to quiet light-enlivened (not necessarily bright, and certainly not dramatic) rooms of a stable proportion.

My office as a room is a silent office, even though we talk there. It is not oppressively empty when it is empty, but peacefully at rest. It is an office more like a church than a factory — and so it should be for I want the work that comes out of it to have something of the same sacredness.

If we think of work as the raising of matter, as provision of food for the human spirit, then places of work need this sort of atmosphere. I think of the old carpenters' shops in the days before they became a screaming tension of dangerous machinery. Not silent rooms, for there were too many interesting things in them, but places of magical reverence.

What sort of a world will our noisy sheds with dead avenues of fiercely powerful machines or cosmetically zippy offices create?

These daily rituals, repeated thousands of times, can have healing effect. Even in places of work, and especially in homes, architecture has the function of providing rest for the soul.

When we come home from a stressful day, the home and the night are for renewal. If they don't provide it, stress builds up on stress and physical or psychological collapse follows. When we go to bed at night we pass into another world and are reborn each morning. How much care and worry can be washed away by sleep! We enter each new day with hope — how otherwise can we survive?

But what haven of calm do we come home to when its inmost sanctum is full of mechanical noise — TV perhaps? How do we bring the nightly renewal of rebirth each morning when we are wakened by an alarm–radio? It's not just people's habits I am talking about, but rooms that *need noise to keep you company.* Many houses, many rooms *need* noise. If we are going to try to provide places where people can live in health, places where people gain rather than lose strength, grow rather than wither, we need to make places where silence can be a welcome guest, where silence can fill the space with its renewing, healing power. This doesn't just mean good sound insulation; it means places of silent quality — to sight, touch, smell and so on — not just noiseless places, but places of healing silence.

12

The Urban Environment

Cities as places

I have met architects who agree with the outlook expressed in this book, but dispute its relevance to the sort of projects they work on. To them, everything discussed so far is impractical idealism. Their questions are: is this applicable to the urban situation? What about the sort of huge and dull projects many architects have to work with? How can one work with commercial clients who are answerable in terms of company profits? Is this way of working expensive and therefore for the privileged only? Can it also be cheap?

Although I have worked and lived in cities, I was brought up in the countryside and the focus of my life has always been there. Even before reading Schumacher I have preferred small, low-impact scale to the grandiose and have difficulty empathizing with people whose activity is motivated by profit. I am therefore somewhat an outsider and as such have a way of looking at things which may differ from an insider's view.

The word 'city' conjures up widely varying images: caverns filled by traffic; intense experience of human activity, both stimulating and stressful; jumbles of decrepit buildings used in ways they were not designed for; clean-edged, bland-surfaced rectanguloids; squalor, poverty and tension as well as luxury and affluence; industry, offices, shops, apartments, people. There are all kinds of cities, and in most cities many kinds of experience. Vegetation-rich, traffic-free, winding brick alleys of human scale may be but a short journey away from multi-storey concrete blocks in bleak grassland deserts or brand-new glass and steel prestige symbols next door to decaying workshops and houses with strong historical associations. In using a general word 'city' to describe this endless variety I am concentrating on the problems and potentials they have in common. Traffic, air-pollution and crime are not the only urban problems. All sorts of factors — social, demographic, logistical, economic, ecological and so on — interweave to create the living, vibrant and imperfect miniature world that is a city. All of these are influenced by its built substance: some of it is beautiful and good to be in, some causes terrible human problems.

Cities need health-giving built environment. They need it very badly, especially as unconscious and invisible processes tend to foster negative, *unhealthy* developments. Already many cities have been distorted to the point that many people experience them as alienating, life-suppressing, hardening, unwelcoming places to be in — but that is not what city life is about!

Whether one likes or dislikes cities it is a fact that they have a central significance in civilization. They *need* not be dominated by commercial and exploitive pressures as they so often are. In theory, cities are the places where civilizing, cultural works are brought together to cross-fertilize each other, where economic activities of service meeting need exist in great diversity with a high degree of symbiotic richness and where there are enough people to support uneconomic benefits for all. If we hold this higher image of what cities *could* be we can see beyond the social, environmental and spiritual disasters that they all too often have become.

Cities are not cities if they do not condense and fertilize the spirit of a region. This does not mean that city architecture has to *copy* local style but that the mood of a region can find expression in the built environment.

When you go from countryside to market town you feel this intensified spirit. You feel it even where towns have been swollen and distorted by industry. Migration from the countryside, still continuing, ensures kinship links between city and region. The huge conurbations no longer respond to their surrounding regions; they dominate them. But Paris, Edinburgh or Washington have the spirit of France, Scotland or America as unmistakably as Bristol or Newcastle have the West or North of England.

Chainstore architects' departments with more allegiance to company image than individuality of place, together with ubiquitous international style 'downtown' imports, de-localize the very heart points of a region — the condensed core of its economic, cultural and social life. Outside these centres, national speculative building companies and standard plan designers with their commitment to economy by repetition continue this process, aided by manufacturers' nationwide marketing.

In too many cities all over the world there are too many 'no-places', those parts of the land which do not invite the passer-by to stop. Some are dominated by through-flowing traffic, some by noise, smell or wind. Many are just too boring to walk around in. Many are threateningly surrounded by boxes to accommodate statistics — so many hundred, or thousand, people; not homes or work or meeting places for individuals and social groups.

Most (though not all) cities and

towns have existed for at least several generations. They have historic roots which live in people's memories: places people knew and used in particular ways in their childhood, places that their grandparents remembered, places that give stability and roots when the boundaries, skylines, streetscapes, usage patterns and local 'atmosphere' change bewilderingly quickly. These are foundations to be built upon. The inevitable evolution of physical environment requirements is faster in cities than elsewhere. Places in which we can see ageing buildings, repair and modification for new purposes, are places with a healthy vigour. Places that remain unchanged except for restoration or maintenance risk becoming moribund.

It is when these forces of vigour cease to provide for needs arising out of the being of a place and start to impose new directions, alien materials, undue physical scale and scale of sensory impact that they threaten to destroy the individuality of a place. When organic growth becomes supplanted by the power of imposed ideas, the spirit of a place is under attack — and it is fragile! In only a few decades many places of soul-warming character and human support have ceased to exist: they have become something else, something with a different spirit, something feeding on and off the more negative sides of the human being — greed, desire, territoriality, mistrust.

Yet urban development and evolution need not be like that. Cities are made up of human activities, the places in them are given *individual-ity* by the colour mixed out of these activities — and there are thousands of colours! A new activity *of the right colour* can add to a place — subject of course to appropriate architectural handling. Indeed, listening to what a place needs is not only beneficial to place but also makes good sense from the commercial point of view.

Unfortunately the modern tendency is to withdraw activity from the street — or even from the window — to be something shut away *inside* buildings. The greengrocer's or ironmonger's street display, the fishmonger's water-sluiced window, the shoemaker seen at work, the dark tunnel into the glowingly-lit fabric shop, the baker's oven smells, the clatter of the printing press just glimpsed within, these are more and more shut away. A whole street block can be just one supermarket. An amazing variety of goods and interesting-to-observe activity goes on within it, but often all you can see are the discount price advertisements in the lower part of the windows and above them receding ranks of fluorescent lights. If we looked *down* into the supermarket so we could see what was going on, if it were arranged in *depth*, not *length* of street, if its public faces were let, or used, for *human-contact activities,* it would have quite another influence on the place it bounds.

Shops — places where material needs and wishes are provided — are central to the experience of living in cities. Attractive, interesting, descriptive window displays sell their products. The broader each individual shop, the slower the rate

With our streets taken over by motor vehicles and our climate and culture conducive to closed walls, we in Northern Europe tend to forget that it is human activity that makes a city alive.

of variety when we walk past and the less life, vigour and varied sensory experience in the street. It is quite a different experience to walk through an alley of little shops, broadening perhaps to a street market, to that of walking between two department stores.

Factors outside our control may make the replacement of small family businesses by chainstore conglomerates inevitable, but by *architectural* arrangements the individuality of each department could be expressed at the building's public face. It is conventional to enter a supermarket at one point, circulate through various departments and pass through a wall of checkout tills to leave. People pro-

gress through a built sequence diagram like materials processed in a factory. Were it possible to enter and leave each department from the street, maximizing departmental individuality without loss of interior interrelationship, the place, customers, staff and just possibly the business itself could benefit.

Approaching architecture with place consciousness rather than building consciousness can to some extent free environment from the constraints of atomistic plot boundaries and insular economic criteria.

Shops, cafés, entertainments and so on bring life to towns, but what about offices? More people spend more time in offices than they do walking amongst shops. In Victorian times factories expressed their varied industrial processes in their purely functional buildings. In contrast to modern factories they were interesting to look at, though local air pollution made them unpleasant to be near. But even in those days, offices tended to be just big blocks with lots of windows — as dull to look at as the work for the clerks inside them.

Of all office building types I personally find town halls amongst the most boring. Their prestigious grandeur and formality seems inappropriate for 'servants of the people': they tend to sterilize a whole area right at the heart of a town. Yet what, beyond large buildings to house administration and bureaucracy, are town halls? They serve presumably necessary functions, are full of people coming and going, and would seem silly if placed on the edge of town, but they are so

The extent to which buildings reveal what is going on inside them can make all the difference to whether a place is interesting and alive or the opposite. This is a matter more of surface treatment than of architectural form. Just add ground texture, a tree and its shadows or a pool and its reflected light and an architectural form that could have been experienced as a prison, now invites entry.

boring to walk past. If instead they were thought of as the regulatory organs of towns they could be seen to perform three functions: administration, which takes place within the building; public contact, which to avoid bureaucratic institutionalism would take place at its face, perhaps also on first floor galleries or terraces; and social pressure balancing. This last, which is after all the *raison d'être* of democratic government, can find forms appropriate to the heart of towns such as immigrant cultural centres, enriching the streets the building bounds.

However boring the precedents, we can look at any projected building this way and ask, 'What is its true *essence*? How can its *positive* contribution to society — all too often buried, in this case under bureaucracy, elsewhere under profit, prestige and so on — be developed?' And we can look at sites as *places*: How can this place be improved? How can the inspired core of the project and the needs of the place

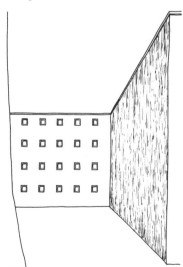

be brought together so that out of their marriage a seed can fertilize a whole district?

Many projects are, however, in their initiator's eyes, purely commercial undertakings. Here, the first requirement is to place the economic criteria in the right context. Such projects often result solely from a desire for profit though fortunately this is not always the case. If they are not going to be profitable, businesses are not interested. But even for those who are only motivated by self-interest, short-term profitability need not be their only motive. Unremitting pursuit of profit is unlikely to foster responsibility, yet profit can be a *consequence* of responsible design. Meaningful staff involvement in office redesign is beginning to be seen as an important element in creating a satisfactory working environment which in turn is more and more recognized as influencing productivity. As staff salaries account for some 85 per cent of average office costs, any improvements which improve health and morale and reduce staff turnover can pay for themselves quite quickly, for the costs of continually training new staff and of the ten-to-five syndrome are considerable. In countries with full employment, employers are beginning to realize that attractive workplace environments are a necessity to get and keep staff. This issue is not limited to indoor environment. Buildings which improve the quality of *places* also benefit their owners by, for instance, making the surrounding area more attractive for staff or customers. In other words an economic case can be made for doing what one believes is right for people and for place.

Looking at the project from the viewpoint of place will show up all sorts of potential benefits — both environmental and economic — that are not obvious when you think of a project as a building. 'Office block' conjures up an image which is quite different from 'street face' or 'courtyard and passage complex'.

Whatever the design, some places will lose more than they gain from some sorts of buildings. With 'place consciousness' these situations are more easily apparent than with 'building consciousness' but it requires material sacrifice to refuse to accept work, especially if one has a large practice to support. The corollary is how can one do something one does not believe in?

My first reaction to many proposed projects is that the place would be better undisturbed. Almost everywhere, however, the charm of a place is due to human activity. We can continue this process of improvement but *if and only if* the development can find a way to take part in the organic process by which places come into being, grow and change. Here the fundamental questions are, is the project founded upon a need in this place, of an appropriate scale, and of a motive that will raise rather than debase the place?

Millions of people live in cities. In many countries more people live in large towns or cities than in small communities or the countryside. Some enjoy living there, some hate it. All aspects of life — stress, work pressures, culture, stimulation and so on — are brought home (like industrial toxins on workers' overalls) to the homes in which people live. Homes have a renewing function: the greater the stresses and inbalances that people bring home, the more acute the need for the home to be a place of healing.

Homes for renewal, for healing from stress, need certain characteristics. The right balance of privacy is one of them. Traditionally, front gardens, yards, basement 'areas' or living half a level up gave such semi-privacy. Steps or bridges across sunken areas were important threshold experiences which, where wheelchair access is a consideration, we can recreate in other ways, perhaps by exploiting the slope of sites or with inset front doors with flower-bed obstructions outside windows.

A real portal of a dark low tunnel leading through into light is both threshold from the outer world and the first step of an important ritual experience. The protective womb-like enclosure of the bedroom is also important. How can one be born afresh each morning from a box?

Some cultures have a whole series of buffer thresholds from the noisy, impersonal and public, to the quiet and private haven: community, street, yard, staircase, home. In Britain there tend to be less, but imagine living in a caravan in a city — you go straight from outside into the living space. You can survive in it, but its renewal characteristics are small. In the country it is not so critical.

Places for inner renewal need space that is not cluttered; rooms need not be large to be spacious. I recently stayed in a smooth box-shaped bedroom, twice the volume of my room at home but half as spacious. Layout, furnishings and light, extent, focus and shape of window, shape of space, texture, colour and vegetation can transform a small room into a spacious calm one or a large room into a poky, busy or oppressive one.

In urban surroundings the home has to resolve greater conflicting requirements and constraints than elsewhere. There is a need for more privacy, but also more spacious, calming, tranquil views; a need for

Even in the densely built-up city it is possible to create havens of peace. To enter this courtyard you must pass through a tunnel under a former factory and turn along a narrow mews.

more light, but not at the price of protectiveness; a need for more indoor, and often private, open space which the price of land makes impossible; a need for quietness and clean fresh air — yet there is always noise.

Between home and the outdoor world lies the home area. Suburbs appeal to families with young children just *for* their quiet streets, lightly enough populated to get to know neighbours. They also have sufficient *private* garden space ideally to 'walk all round my house'. It is just their tameness and low density which denies them convenient public transport, that makes them so claustrophobic to teenagers. They need a 'home area' differentiated in density and activity and alive for longer periods than the hours of sleep.

If there are too many animals on limited territory, diseases and social aberration appear. Studies seeking to relate these findings to human communities have tended to founder in the complexity of human life.[1] Crowding in different cultures, air quality or weather or noise climates cannot easily be simply scaled and quantified. None the less there is a widespread view that stress is proportional to crowding. This may indeed be true, but attempts to relieve stress by reduction of city densities have led, in the Californian extreme, to situations where the roads and car parks that necessarily result cause more noise and air pollution and higher machine to human space ratios than

[1] Ian McHarg, *Design with Nature*, Doubleday/Natural History Press, New York, 1971, pp.187-195.

had urban densities been conventionally high. To city and regional planners the lesson is one of scale and decentralization; more locally, the issue is how to maximize the advantages of a lot of people living close together while minimizing the adverse effects.

Density can be thought of as a stress or slum factor or just a measure of how many people can be packed onto a particular site. If, however, we look at density figures as a contour map of human activity — with its attendant noise, generated traffic, lack of space and sunlight, requirements for paved ground surface, along with shops, services and public transport — we can get a picture of what uses, atmospheres and densities are appropriate for each other.

Like commercial developments, but in different ways, the effects of housing extend far beyond site boundaries. Whatever the level of involvement individuals may have in the design of their homes and associated communal places the surroundings outside these protected, sometimes even exclusive, enclaves are shaped by other factors. Many of these are driven by profit-making motives. If these factors dominate, they can pollute the spirit of a city; if they can be countered by environments which meet needs of soul rather than desires, there is still hope that civilization will prevail over its pecuniary double.

I have not yet been to a city in which, either for humanistic or profit reasons, areas of old residential buildings have not been demolished and replaced by new. Experience has taught us how damaging this approach has been to neighbourliness and community and that physical problems of slums, such as mould growth and structural deterioration, can reappear as a result of the mismatch of buildings to users' lifestyles. What the demolition ball smashes is not just buildings but that web made up of personal relationships, memories and thoughts which, together with what has been growing over a long period in a place, makes the spirit of community, of place. Hopefully the era of urban redevelopment by demolition is behind us, but there are other ways in which sectarian self-interests threaten places.

Every four years a new city wins the opportunity to host the Olympic Games. It thereby gains prestige, an immense inpouring of money and at the same time stimulates a ferment of demolition, building, motorway and parking construction. The price of the economic benefit leaves a lasting mark on a whole region. This sort of thing is happening in miniature all the time: the more money available for investment in money-making projects, the worse the damage.

Britain suffered a similar period in the Sixties where even in small towns market squares were effectively turned into car parks — and that was before motorized shopping turned everything around, focusing shopping upon the through motorway in place of the meeting place — the town centre where separate roads come to a single destination. This sort of development has not only been at the price of character but also of place. In many cases places that were the coalescence of

community spirit are now no more than names on a map. Some have been obliterated by the pressure of traffic, which starts by turning a shopping street *focus* into a barrier, a *boundary*. Life withers at this boundary which advances as roads become wider, as blight becomes more intense and extensive. Other places have been physically obliterated by demolition, becoming merely sites for new — often well-intentioned — redevelopment. Others again have become dominated by new buildings, uses and users which change their character and focus beyond recognition.

Can such places, rich as they are in history living in inhabitants' memories, be refound? How can they again become living magnets, places to enjoy, to love and care for?

Before anything else, they must be places that invite one to stop in, not just pass through. In one way or another, this means solving the problems of through traffic pressure, for instance by slowing it down, by re-routing it into existing 'boundary' lines. I used to live outside a small seaside town. Over the years the road leading to it grew from sleepy and winding to a four-lane trunk road, cut through the landscape. They led to a densely built old town with minimal car parks, on a headland surrounded by sea! Roads are very expensive to build, car parks destroy towns. I wonder how much cheaper it would have been to provide parking out of town (on fields perhaps, because the pressure only existed in fine summer weather) and a free bus service into the town.

Some traffic problems can be

The high density juxtaposition of different activities and individuals is at the heart of urban living. Such places are attractive to visit but, to those who live in them, inadequate sunlight, lack of play space and green, proximity of street noise and the claustrophobia of living cramped by neighbours can be oppressive. At certain stages of one's life, the social and cultural richness born of proximity outweigh the disadvantages. For families, however, we need to find ways of combining the density of urban experience with the renewing qualities of light, air, greenery and space to feel free in.

ameliorated by rethinking how people travel and by subsidizing public transport, but whatever the attempts, traffic problems do not go away! Sometimes places need to turn a strong back to traffic; a back that excludes its noise, fume, speed and risk. Such walls or cuttings

benefit by design with absorbent surfaces, plants and acoustic section.

It is not always however desirable to turn a back on outsiders, even those who exploitively pass through. Where traffic is slow and easy to cross is the place to see and enter through protective-edge

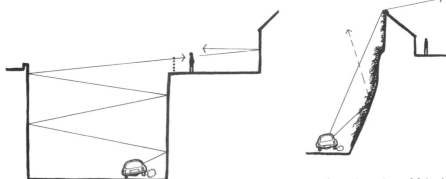

Acoustic sections: Noise is intensified by echoes from buildings or cutting walls (left). Inclined surfaces of absorbent materials can reduce this effect (above).

barriers.

People on the streets are not just stationary but always doing something, going from one place to another. You need to have a reason to stop! It isn't just a matter of having the right facilities to draw people to, it is how they are arranged and supported by outdoor atmosphere that will determine whether a place can become a magnetic focus for an area or just a shopping, entertainment, business or workshop zone.

Life is speeding up. You have only to read an old novel to notice how much faster a pace we have got used to. A characteristic of over-speeding, whether at the car wheel, at work or in daily living, is that we don't notice it. Stimulation increases until it crosses the threshold of stress. Some people think that the number of psychiatrists in New York is in direct ratio to the pace of life — a pace immediately visible to outsiders even in the speed at which people walk. Many people need holidays in slow-moving places for their health. All of us benefit when places give inducement to stop. The development of *places* as distinct from *routes* doesn't just make cities look nicer but is fundamental for the health of their inhabitants.

The evening-out of all sorts of experience, from advertising intensity (however harmful it is), building scale, traffic speed, spatial characteristics and so on water down the experience of where places start and stop. Big cities were, and to some extent still are, made up of little places. Sociability, founded upon re-meeting the same people in different circumstances, is

much easier in small communities than large. Locality and identity can depend more on social and sensory character, colour, scale and spatial characteristics than on shape or style of buildings. An analysis of randomly directly black and white photographs will show how much stylistic variation can exist in an apparently uniform area. Materials, constructional and spatial principles, however, are usually more consistent. Those are only the visual characteristics. Important as they are, identity is much more than that: activity can be measured in gradients. There are both magnetic foci and steps at certain thresholds. Thresholds underline the transition from one place to another, make *large areas* into a pattern of *smaller places*.

Thresholds can be abrupt changes in urban character, like two-storey buildings around a five-storey corner or different tempos of life, or they can be portals. To pass by foot through a narrowing in the street, with different textures, lighting, uses, architectural scale, beyond, tells you unmistakeably that you are entering a different place. So does an archway, tunnel, steps or gate.

Gated estates have made their appearance in America and their object is to exclude trespassers. In Britain there is a move to close pedestrian through routes. After all strangers might be coming to rob! But when I am a stranger in a city I don't want to feel like a trespasser. I want to experience the differentness of the place and walk wherever it is interesting.

Crime and private property are growing phenomena. They feed

Through arched passages you can enter another realm: an unexpected courtyard, a quiet street, a residential area. The scale, darkness, texture, shape, slope and twist of the passage can help reinforce the change of inner state that accompanies the change in outer place.

each other. There are two broad schools of thought about architectural means of reducing opportunist crime. One is to enhance *private* realms so that every place is guarded by its private owners. The other is to develop *public* life so that places are always full of life and never abandoned. Both have social and political implications. Personally I do not doubt that the current exaggerated advocacy of the former will, locally, result in lowering robbery statistics. However, in its stress upon the *privateness* of property it is feeding a tendency to take for yourself whenever you can. In society at large, therefore, it undermines civilized values and may well be counter-productive. It is, after all, in the most privately owned country in the world that it is least safe to walk down the street.

If we add Steiner's observation on the relation between aesthetics and moral development we can see that public spaces need to be places of attractive identity, of spirit-nurturing spatial, architectural and sensory quality infused with the 'colours' of human life — places of spirit. When places are improved there is a tendency for them to be taken over by a new wave of middle-class occupants. The very success of environmental improvement risks failure in the social sphere due to the problems of differential buying power. City areas experience the same pressures that turn Welsh communities into English holiday houses, empty most of the year and unsupporting the community life.

So long as success can be turned into monetary value, this risk remains. Where improvements can be related to appropriate tenancy structures, levies on resale price increases and the like, or where they can have community-specific characteristics, there is a chance of reducing 'middle-classism'. Areas designed for a local-contacts network centred around workplace, pubs, facilities and children's street-play, with restricted provision for residents' parking, would have more appeal to people whose lives were used to this pattern then to those whose employment, entertainment and social patterns were less local.

Whether or not we bother about profits people make out of it, environmental quality is interwoven with economic cause and effect. There is a view that investment in money-making activities will set in motion a spiral of environmental improvement. To attract the right people and get the ball rolling,

depressed areas need a new image. They need environmental improvements to stimulate the economic recovery that will *really* improve the environment. But what sorts of improvement? And who gets pushed out?

I look at things the other way round. Awful places demoralize people and encourage their dependence on bought supports. They depend more and more upon money — and it isn't always there! Human places are not immune from these problems, but they can support a greater resilience — and also a better self-respect as the foundation of will. If qualitative improvements lead to economic benefits it is less because people make money out of art than a symptom of a state of social health.

Cities for people

I have discussed making *parts* of cities into *places*, an essential step in bringing health. But to live in a city is different to living outside one. People need different emphasis in their environment. But what *do* people living in cities need? It is easy to talk vaguely about other people, but what would I need if I lived in a city?

Foremost must be sociability. Cities can be lonely places where informal meetings are strangled by alienated inhibition. Where people can meet depends to a large extent on the provision of appropriate facilities — especially places where people *do* things together. Doing things together, such as childcare, work, study, sport or participatory entertainments — particularly with the same small community of people — breaks down barriers.

More and more these days, people move home. Economic and fashion currents can concentrate this into trends where newcomers can be readily classified as different. Newcomers and established residents can resentfully see in each other representatives of an alien group, not individuals. Once you get to know people, this is reversed. They are individuals foremost; differences in race, occupation and background merely make them more interesting. To quite a large extent *how* people meet is supported or hindered by the environment. If I wanted to hinder this community-building process I would design urban flats where the only places you come into casual contact with other people are concrete access balconies, impersonal corridors, lifts and refuse chutes, none of them places conducive to social contact. For the better off, separate housing units with private, self-contained gardens and en-suite garages may be highly desirable, but if you drive to work, to the shops, to friends and entertainments, how do you meet your neighbours?

Without any loss of essential privacy we can encourage activities to overlap in the communal realm. To have suburban shops within walking distance requires not only site allocation and appropriate housing density but also a network of short-cut walkways that are inviting enough to wean people off motorized habits. Many urban housing projects have village-sized populations, but as they are housing projects, they are for housing only; everything else is just round the corner. Here we need to look at every activity and amenity to see whether it can have a social-building function.

A community laundrette can be a sterile, soap-smelling fluorescent-lit hole that you spend the minimum time in or, especially if it has facilities in which you can do more than press buttons, a place to make tea and wait till the wash is finished. If it is not inviting it won't help to make friends any more than will the parents' bench by the sand pit if it is concrete, in draughty shade and ugly, noisy, exposed surroundings. How many empty park benches have I seen on landscaped roundabouts! Daily life is full of little activities that, if in welcoming surroundings, can support meetings between strangers.

I doubt if anything contributes so much to social malaise as anonymity — the feeling that you know nobody and nobody cares whether they know you. It is certainly much easier to shop-lift a department store than a shop where you know the owner. In this respect, if no other, the qualitative characteristics of environment can work for or against social health.

Urban vitality and sociability are not enough for teenagers. They need to break out from the protective claustrophobia of home life and find adventure in the real world. It can be morally inspired adventure; protest movements offer such opportunities. It can be existential, exposure to a flood of powerful new experiences. In the super-stimulated time we live in, existential adventure can be much more dangerous to the personality than it seems. You can't protect people from themselves, but just as adventure playgrounds are designed to feel more dangerous than they are in contrast to playing on the street — which seems tame until there is a traffic accident — it is possible to provide an environment which maximizes the excitement but not the risk.

Adolescence is the period of growing towards one's personal identity, of becoming aware of oneself as separate from family, from institutional groups (such as schools), from others. It is marked by a high level of peer-group conformity as protection against the insecurity and loneliness this separateness opens up. The entertainment and clothing industries exploit this conformity as marketable fashion, changing fast to stimulate new sales. Whatever its ills, environment provided for young people competing on the customer market — like youth clubs — needs likewise to have its ear to the fashion of the moment. As there is always a big time-lag between planning and completion, it is questionable to what extent environments 'provided for' can avoid being seen as philanthropic institutionalism unless they can be adequately adapted by the users. Flexible adaptation does not mean the provision of empty sheds but of places with varied qualities which can be enhanced, styled and used as required and which young people *themselves* can build, adapt, furnish and so on.

This process of becoming aware of oneself is also a time of becoming aware of one's desires. A lot of money is made from catering for desires, but the characteristic of desires is that they can never be satisfied. You always need more. . . and more. A society ruled by desires could not be sicker.

It is the transformation of desire from masters to servants, from sex to love, from personal status to inner resilience of character, from craving to independent discernment that marks the development of an inwardly free individuality. Market forces feed the former state, extending the enslavement of desire back into childhood and far into adulthood. You only need to look at toy and car advertisements to see this.

To design, therefore, for what young people *need* is to transform what they *want*, to provide environments where they meet more as humans than as sex-objects, working relationships where mutual responsibilities are more meaningful than prestige and where socially inspiring necessity calls forth self-sacrifice, hardship and adventure; in other words to strengthen the development of the individuality and its mastery over the realm of desire. In the age of glue-sniffing, crowd violence and bitter divisions of wealth and opportunity, society depends upon such individual transformations.

It is easy to see where projects which are *inspiring*, such as work with the community, environment or the arts, can play their part, but what about architecture? Especially, what about architecture *not* associated with inspiring projects?

Above all, we need buildings and places *welcoming* to the soul: places that are not exploitive, places that, in the way they are conceived, planned and built show love — that most needed and least supplied quality — that can transform the social delinquent into a crusading rebel or the competitive success figure into the servant of a great cause. Easy? Obvious? Then why doesn't it happen more often?

Then we need understanding. I was recently approached to design a hostel for urban teenagers in the country in a site selected for its spectacularly beautiful surroundings. I was asked to provide a social room with the atmosphere of a pub (without alcohol), a night-time place, yet located where the views and sunlight were best. I questioned whether the urban-consumer attitude these teenagers apparently would bring with them was appropriate to the objectives of the project. Should the environment instead provide support for creative artistic participatory activities? Group tea-making instead of a bar to buy manufactured soft drinks in disposable cans, live music instead of a juke box, a cooking, baking, open fireplace instead of a purely amenity one, sunset and candle light instead of a dark-walled electrically-lit cellar atmosphere? The project was postponed (or perhaps they got someone else to do it) so I cannot say whether my suggestions would have led to a mass exodus to the pub each evening or to a waiting list to stay there.

In the competitive city environment youth clubs have to compete with commercial establishments. They need, as I mentioned, to be at least in step with possibly fast-changing fashions. But the super-stimulation of commercial clubs with their extremes of light, colour and tempo can destroy the sense of wonder, that most valuable and fragile of human faculties. Without wonder nothing is new except the dangerous, nothing is inspiring, nothing worth putting one's own interests aside for. Like 'super-sell', 'super-stimulation' presses upon individual freedom. If on the other hand we can maximize the 'soul-colour' of places, we can support the sense of wonder. For this we depend upon the full sensory palette of architecture to make different qualities to colour, enhance and encourage different 'activity-colours'.

Mid-childhood is the age of dens and wilderness play. As urban wasteland becomes more dangerous due to demolition work or guard dogs, this frustrated creative play can find more and more socially unacceptable forms on the street — supermarket trolley races, mischievous damage, breaking trees and so on. There is plenty of need for chosen, manicured landscaping, but children need something they can adapt, something more of a wilderness garden with places they cannot be seen in, branches and stones build things with, water they can dam and divert, steep slopes to run, roll and slide down, opportunities to scare themselves without serious danger. If these places aren't there, the streets are a ready magnet. In quality, these adventurous places should not be too hard, dirty, aggressive, loud, for they have a very formative influence.

When children draw their first maps they start and end with home.

Home and the world outside it are their whole world and they will bring the qualities they meet there into themselves as they do the qualities in people that they observe. Here above all environment must not be too hard, aggressive or dishonest; rigid forms and spaces have an entrapping, sterilizing effect on development which produces antisocial outer activity. Mobility of shape encourages inner mobility. Harsh, colliding hard-edged rectan-

A swing, climbing frame or rope bridge above a slope feels more dangerous because you look down a long way. A fall however would be no further than onto flat ground. Whenever we design for adventure play — and this is everywhere children are, not just adventure playgrounds — we need to consider ways of maximizing apparent dangers, reducing real ones and, in so far as we are able, eliminating unseen ones.

Below *Small children drink in everything in their surroundings, both animate and inanimate. Everything they experience is reflected in their state of being; indeed much of it is imitated in play with penetrating accuracy. Harsh, immobile, imagination-suppressing surroundings are hardening and damaging to children's inner growth. They need places soft and fluid enough and wonder-filled for their imaginative world to blossom (unfinished play alcove, with coloured glass windows, for four to six-year-olds).*

gular forms, uninviting textures, unresponsive ground surfaces, conflicting sensory information and the like have a hardening, distorting, stultifying influence on the developing personality. If we blame modern harsh architecture for opportunist crime, think how it must be to grow up in it!

It is not just growing up not knowing that milk, cows and grass have anything to do with each other, not knowing where the moon or even sun will be, not varying play according to season, never hearing silence, perhaps never experiencing wonder; it is also to grow up in an environment with a high level of aggression to the senses and sensitivities. But we can do something about it. We can design places that enjoy, savour the different qualities of sunlight, places which vary significantly from season to season — choosing for instance shrubs and trees that are in flower and leaf for shorter rather than longer periods to maximize seasonal progression through different plant varieties. The breeze can stir leaves; different leaves move in different ways. Rain, bringing boosts in plant growth, magic reflections and, to children, puddles to splash in, can be something to enjoy *if* we live in places that can draw out these benefits. Elsewhere, by the exposed bus-stop, in spray driven by traffic or by high turbulence, on the mud-squelched playing field or in any damp grey place, it can be a misery.

Conscious choice of materials, scale, appropriately-shaped spaces and boundaries of forms, poetic conversation between elements and between successive environmental experiences — some metamorphic, some contrasting, some unexpected — are not just good architecture but health-giving, formative influences on the growing child. What childhood experiences do you remember of urban places? For me it is the smells, textures, sounds and space of different streets. I remember the long straight ones and the more interesting curved ones. I remember the places that were sensorily and specially more alive and the dull dead ones.

Growing children need life-rich food. They do not grow up healthy on deprived diets. Nutrition does not only enter the human being by the mouth but through all the senses. The importance of living lines instead of dead, mobile forms and spaces instead of rigid, metamorphosis and conversation instead of repetition and imposition is too great to be left to profitability criteria.

But also in cities live the dispossessed and under-privileged, unemployed, materially poor, even homeless and social outcasts. To a greater or lesser extent public conscience in the form of state or voluntary agencies focuses on material poverty, providing essential material amenities. There is no money left over to 'beautify'; but it costs no more to imprint spirit into matter than to build containers for statistics. Love does not cost money, nor does user-involvement in design and construction — in fact things are cheaper that way!

It is in the poorer areas of cities that we find the worst environmental conditions — and naturally so, because the better-off buy their way out of these districts! These are the places where the environmental, architectural, social and personal problems of cities are usually most acutely visible. Without monetary buffers the poor are exposed to the life-sapping qualities of the everyday built environment. How many of us experience song in the heart when we walk through mass housing projects? And we live by the heart!

Places built for poor people need to be cheap; local authorities have only so much money. But cheapness is no excuse for poverty of environment. Cheap or even virtually free materials can be the most attractive, approachable and alive. Poor districts of older cities are often graced by ivy, plants and moss on walls, roofs and paving. Build-

Hand-made textures bring so much more life to a place. They are something more easily afforded by the poor than by the better off.

ings are carefully patched, not restored; they age gracefully. Construction is often of stone, brick, wood or finished with whitewashed or coloured render. These *materials* are rarely significantly more expensive than the accepted norm: some, like the vegetation, cost nothing, but they all cost time to construct and to maintain. Where labour is free, time is freed from its shackling equation with money.

When the building process involves people as human beings and not as mechanical repetitive producers, buildings tend to cost more. These cost increases can be offset by less status-orientated materials, components and luxury equipment, but such buildings can never compete cost-wise with mass-produced dressed-up boxes. On the other hand, when people build them themselves, even only in part, their buildings can be much cheaper than anything from any producer's system. Where tenants

It is in the poorer parts of cities that 'weed' trees grow up on abandoned land, yet it is they that make the surroundings, however dull or lifeless, bearable.

and other occupants can be involved in construction and maintenance they are free to build environments of love — which richer districts are not. This can only happen with people who are inspired and this in turn depends upon a listening design process in place of an imposed one, a process whereby people's needs of soul can find a form that they can work on to build the places they deserve.

Individual design can be a licence for an individualistic free-for-all. But if you can plot a design sequence from overall community atmosphere through local area quality, communally-responsible building exteriors to family-controlled interiors, the design process can be

a first step in community building.

Cities are for all: children as well as adults, poor as well as rich, yet the decisions that shape them are largely made by a restricted social, income and age group, mostly men. All actions affect other people: we can either not think about the effects, categorize the other people simplistically or try to project ourselves into their state of inner growth. Only from this last approach can we ask questions as to what will support these (unknown) individuals and what hinder them? If we rely upon outer analysis we can do no more than react to what already is, what has come into being through the disproportionate influence of the better-off. Reaction causes counter reaction, rapidly moving towards the political sphere with its bitter divisions. To heal we need insight, and insight means stepping beyond our own narrow boundaries. Real insight, because it is concerned with the human being, is always spiritual.

Cities for life

Although they bear the worst brunt of the worst surroundings, the problems of cities are not confined to the poor, to minorities, to children or to any other under-represented groups. A major *personal* problem is the loss of context to personal identity. Large communities with artificial means of life-support have lost their 'translucency'.[2] Unlike village life, we cannot easily see the systems —

[2] A concept I owe to Leopold Köhr, the father of small-scale socio-economics.

social, ecological, economic — that flow through cities. Daily personal encounters are not part of a social pattern; urban growth is not a visible organic process but a series of unasked-for impositions unrelated to previous social form; systems of food, waste, water, mechanical energy flows and so on are complex and invisible; topography, watershed, ecological and climatic individualities cannot be recognized; seasonal and even diurnal rhythms are greatly reduced in their impact. In cities I have often had to think what was the time of day or month of the year for I only experienced day and night, summer and winter, with perhaps just a hint of dawn and sunset, spring and autumn.

Vital as they are, it is not just a matter of ecologically healthy supply and waste cycles. If we can experience growth and change as an organically developing process it can help to root us in time and place. The more we can experience the working of the living on the enduring, so that it is set into life yet always there, the more this rooting is enhanced. This means, for instance, design in which sunlight at different times of day and year creates significantly different moods, landscaping which makes visible the progression of the seasons (as distinct from plants which stay unnaturally long in leaf, flower or berry). These rhythmic changes play over a durable topography. This, even though visible as sloping ground and sensed as microclimatic variation, is hard to grasp when you cannot see it as a whole. We need to re-create a coherent sense of the shape of the earth

beneath our feet. That these things give a rooting security to people is easy to observe in small communities. We can also work with this health-giving theme in larger communities.

Most city dwellers are denied contact with the food cycle: food is something packaged; farms near cities suffer from vandalism, both unintentional and deliberate, but allotments could well be more in the public view. Farmers' markets with their seasonal rhythms of unprocessed produce bring a stronger sense of season than do supermarket Christmas advertising displays. Rats may make urban compost heaps undesirable but the act of separating compostable and non-compostable material raises consciousness.

How many people know where water comes from and goes to beyond the limits of tap and plughole? Used water could be cleaned of pathogens, excessively available nutrients and even chemical pollutants by a biological and rhythmic flow system of flow-forms, ponds, reed beds and other vegetation (as described in Chapter 7). Such systems don't need to be shut away in sewage farms; they can be attractive, even artistic.

In nature, even in farmland, there are marked differences in microclimate in different topographic, moisture-rich and sun-orientated situations. They are essential to the full experience of a landscape: the windy ridge, the sun-drenched slope, the sleepy hollow. So central to the design of landscapes for pleasure or production, why are they so rarely considered in city

Rainwater — millions of gallons — disappears instantly, often causing flooding downriver. Yet it could course through streets and courtyards in broad shallow streams, drawing bickering indoor children to play in the open air, making rainy days something for them to look forward to.

design?

City life depends upon mechanical support. This grows exponentially with size and speed of life — and these feed each other. Cars, lorries and building machinery make a big contribution to noise. It is hard to find silence anywhere in the world, but I am always struck when I visit cities that is is *never quiet*. We soon come not to notice background noise but it works deeply on us. Quiet is so strikingly therapeutic, think what noise must be doing all the time.

Anything which reduces noise improves urban environment — especially those reductions which shift the focus from mechanical to human sounds. Public transport, low-speed water transport, cycle-ways — preferably outside the exhaust zone — and especially, walking routes, all do this.

When I think of European cities, I think of trams, cycle routes, water buses and networks of passages and stairs. Impoverished in this respect as British cities are, at least you can walk in them. In many parts of the world, especially the USA, it is not considered safe — indeed in some places it is not even possible. Lamentably, city walking nowadays involves issues of safety from assault.

Assaults tend to happen when there is nobody else around. We need to think therefore in terms of continual casual observation. Sight lines are straight, but straight-line streets and alleys are uninviting to walk along, so self-defeating from the safety point of view. We can think therefore of magnet activities during risk hours ranging from late-shops to youth clubs to keep the danger-spots populated. Where particular risk locations can be identified we can locate observation activities; sheltered housing, cafés and upper-storey restaurants for instance are situations where people tend to like to look out of windows. The greater the mix of activities, the longer the life-hours of public spaces and the more interesting they are to look at and to walk in.

One response to growing assault rates is to drive not walk, but driving causes pollution, noise and demands on space. It compounds urban problems and can be stressful and slow. In rush hours when I lived in London I often found walking almost as fast as mechanical travel. Some routes, however, had to be along main roads where all the calm of walking alongside canals in the dark, or the interest of narrow alleys with delivery-men manhandling goods was obliterated by the noise, smell and mechanical aggression of vehicles.

Some cities are well supplied with walkways, pedestrian, railway or river bridges, sometimes stairs, alleys, cliff-edge-like footpaths or arched tunnels under buildings — experiences I choose to include on any walking route. Unlike concrete bridges between multi-storey blocks they are not associated with

street crime, the alienation of children from street play or parental stress. On the contrary, if they are of appropriate materials, colours and textures (under foot as well as on the wall), small in scale, varied as one passes along and never quite straight, they can rejuvenate instead of stress. Straight passages, especially if reinforced by symmetry, force one in a particular direction; turns and offsets bring a sense of freedom and of life.

In other places, quiet cul-de-sacs have no exit for the pedestrian. Often only a wall, derelict building or corner of an under-used storage

In one project I was asked to advise on, one of the site's few visual assets was a narrow brick-walled alley which, however, had a record of sexual assaults. We proposed that the social centre, which the would-be occupants wanted, be opened to the public as a café to extend its social and financial viability and draw more people over longer hours. Placed at first-floor level, overhanging the entrance to the alley, both window observation and people on the pavement would bring this former danger-spot into public view. As a byproduct, we could also create threshold portals to the residential zone and the footpath.

Derelict land, available for co-operative self-built housing for young people.

Social centre

Alley

High walls

Residential street

In the semi-protected world of an arcade, the focus of attention and sensory climate can be more human-orientated than in the busy street.

yard is the only barrier. Civic investment — minuscule compared to road-building — could open up walkways of which only fragments are lacking.

Were there short cuts, routes which avoided through-traffic, raised pavements, colonnades (especially with their repetition of enclosing curved shapes to give protection from the road and focus attention on human trade), slower speed limits (30 kilometres per hour in parts of Europe) and pedestrian priority at traffic light crossings, walking would be more appealing and faster. This in turn would contribute to better air, less noise and fewer economic burdens on society as well as the obvious benefits to individual health.

But walking isn't only rejuvenat-ing — it can also be life-sapping. To walk through some cities is to walk on even pavements through even activities in an even texture of architectural environments, punctuated only by main roads. In others, concentrated activity, where human relationships with strangers become dominated by competition, rises to crowding stress. Sometimes the contrast between commercial faces and stark sterile parking and loading areas conveys exploitationist motives. In too many cities the economic forces behind entrepreneurial vigour dominate the environment. Huge financial and administrative institutions condense the monetary aspect of economics — at heart the sphere in which people work together for each other's benefit —into imposing buildings with forceful object image and lifeless blank faces, starving the passer-by of living experience. Nowadays some are even mirror-faced. To get

Even a three-step (60 cm) raised pavement, with railing, makes a considerable reduction to the sensory assault pedestrians experience from traffic. When, as here, the pavement is at first floor level, noise, fumes and visual impact of traffic are noticeably less and our experience is more of people, displayed goods and place than of vehicles. Devices like this, together with arcades and passages, can increase the level of human activity, and therefore urban stimulation we experience, while reducing the stress-inducing qualities usually associated with increasing population density.

173

a feeling of what is going on inside is like trying to read the thoughts of someone wearing mirrored sunglasses.

Other places are given over to super-sell — exploitative pressure, which does not leave the individual free. It's easy to blame shop managers for any excess but I notice that often their displays are quite acceptably low-pressure. It is the architecture which forces focus upon their wares. Much depends upon scale: short alleys, streets or arcades are routes which happen to be lined by shops, longer ones are shopping *zones*. You enter an environment not to taste freely but to buy! Arcades and malls remove the sky and living light, focusing the environment on its commercial function. Sometimes even shop thresholds are removed, making the goods on display seem unowned, unguarded, there to take. Whilst street-crime is blamed on architecture that does not define territorial boundaries, shoplifting is encouraged by such an arrangement. I once knew a schrizophrenic patient who, when she went shopping in quite a modest market town, saw only witches and devils. When I first went into a new large mall — a town within a town — I experienced in the environment what she saw in the people.

These are cities of hard, dead forms, dead straight lines, unapproachable materials, even and unappealing sensory experience — cities which are not life-nurturing, where stimulus is supplied as a saleable commodity, something to become dependent upon. In these environments money is the life-support; they are not places to be poor in.

Yet shops, commerce and industry are the life-blood of cities, as is the traffic which serves them. It is stopping of the traffic which caused places to start to grow. In days when it was more human-related, traffic brought into being places to stop, buy and sell. Iron wheels on cobbles, a horse for every horse-power, did not make such streets less crowded, noisy and smelly than they are today, but street life was coloured by people, not fast-moving pressed-metal containers. Conscious transport policies can — to some extent — re-apportion focus from alien mechanical things to *people*. In the same way it is possible to look at production, office work and trade in such a way that human aspects can contribute.

The organic development of places creates situations that invite shops and entertainments to grow up. Locations appear which would well suit offices, but not much else. Such a process requires listening, but out of it can grow quite another sort of city: perhaps narrow shop- and service-workshop-lined pedestrian streets, which with steps up slopes[3] and side alleys are short-cut routes and therefore always busy; larger workshops making visible the activities upon which the economic life of the district has

[3] In seeking to maximize the interest in changes of level I do not wish to penalize disabled users. For every short cut via steps there must be an alternative sloping ground. Occasionally we may want steeply-sloping squares or open spaces. Diagonal routes across these can afford wheelchair or pram access.

When we look at old photographs of places we know well we can often see how little has changed. Yet everything has changed. What were then sociable places, bounded by buildings, hedges and so on, are now linear — and therefore unsociable — corridors of motorized movement. Squeezed to the side are narrow bands for pedestrian movement.

Places which once were coloured by human meetings are now compartmentalized strips, dominated by the noise, speed and latent aggression of the motor vehicle. Yet the actual material alterations are small and subtle — in many cases too small to show up on a 1:100 scale plan.

Typically, when towns have become adapted for vehicles, large parts are dominated by traffic. Often 80 per cent of the visible space in which we move is flat road surface and aggressive unrelieved horizontal movements, the buildings rising vertically to contain our vision. Towns for people traditionally enjoy a much more lively interplay between horizontal and vertical.

grown; offices arranged not in large blocks sub-let into small units but 'personal-faced' around peaceful courtyards. Entrepreneurs regard this as unprofitable, but if they rented offices in the Inns of Court in London they might think otherwise!

These might perhaps find form as flowing streets, twisting alleys and stairs, open, closed and sloping collonades, non-square squares, visible human activity — on the street, in open workshops or seen through windows — at all scales from watch repairs to dockside cranes. Places can come into being where sociable commercial functions organically grow up — quays, streets and larger squares where it is natural for all sorts of commercial interchange to take place; off-street courts and first floor pavements and passageways where offices have individuality without sucking life out of the street. Such an approach gives opportunity to adjust 'perceived density' so that small commercial centres can be busy enough and large ones can be diffused into a less pressured network of small-scale alleys and squares.

It is all very well to think of quays as nodes of activity, as key meeting points between sea and land-orientated activities, but nowadays the empty docks are dead ends, leisure centres or exclusive enclaves compartmentalized in time and social group from everyday life. Must they be? New approaches like the water bus and water-borne goods delivery can reactivate them as centres of activity. In the same way that we can think of the social activity aspect of what conventionally are dull, faceless blocks of offices, so we can look at dead places and try to see what catalysts they need to bring human life and energy back to them. Even the deadest of city places has the potential to be like those cities made up of life, cities with inviting walking routes and inviting stopping places, each place condensing a 'colour' out of the activities that bound it. These are cities of many living 'colours'.

Concentration of visible human activity to create life-filled places is central to the 'city experience'. Yet for some people and in some circumstances stimulation can cross that undefinable border into stress. Curves and gentle lines, plant-covered walls, approachable materials, varied and softer textures underfoot all help relieve these pressures. But we also need to provide havens where the soul can rest: quiet places, sun-filled and dapple-shaded walled gardens, entered deliberately through an archway, perhaps at a different level from the footpath; places near water — fountain water, flowing water, calm water, pools, rivers, canals, large expanses. Water is the greatest of all calmers; almost every city has grown up around water, its commerce, power or bridges having brought it into being. Yet nowadays all too often the riverside is the place for a motorway, for car parks, warehouses, inaccessible industrial zones which are no longer even dependent upon water except perhaps to pump waste into. The small streams are now all underground sewers and their loss is a major factor in obscuring topographic understanding. Yet in some cities streams crossed by many little bridges rush beside the streets bringing life to the air.

All too often, however, cities — but not their children — turn their back on ponds, canals and streams. Out of adult sight they become not only squalid but dangerous; some cities fill their docks with rubbish. Water is perhaps the greatest environmental asset any city has, yet how rarely it is developed!

Sky — its space and quiet clouds — is a healer that is everywhere.

Visible sky, natural light and sunlight all powerfully influence mood. Many people who work indoors crave sunlight to the extent that they spend their holidays getting it. Insufficient light can be linked with depression and suicide statistics. Daylight, as I have described, is life-bringing. Dead, grey, polluted air induces a similar grey mood in people which, when we take the morale factor in every kind of activity into account, makes as strong an economic argument for pollution control as does the cost of corrosion of building fabric. Fortunately even in the most built-up areas there is some sky. What would the cost of domes over cities be in terms of additional psychiatric hospitals?

The less sky there is visible, the more trapped, stressed and depressed people tend to become. If on the other hand buildings are low enough or widely spaced, urbanism rapidly drifts into suburbanism. Cities on hillsides benefit here — they can enjoy squares ending in balustrades, steep alleys pointing out into space, unexpected galleries, staircases and ramps in the view and wind, amongst the treetops or looking over rooftops. Selective openings of space to *focus* attention on skyscape and distant views are more effective than general space allowance which sometimes just makes for bleak deserts between cliffs of buildings.

Close-walled streets and courtyards, often necessary to achieve an urban compression of experience, make for dark rooms at ground level. These are the place for indoor-orientated or night-mood activities; pubs, restaurants and

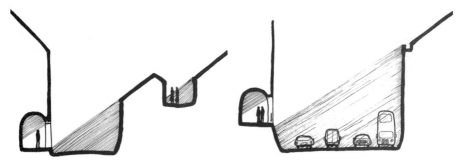

small theatres grow up in such places. Daylight at street level depends upon reflection off materials (including the ground) as well as the amount of visible sky. Damp concrete paving can make the whole world seem grey; brick can make it all seem warm. Whitewash can brighten a north-facing yard, but paints need careful use. Singing, light-bearing colours especially can be sullied by grime whereas others are more tolerant.

How many square miles are covered by buildings which try to avoid sunlight? Painters seek north light so as not to distort their colours, whereas office and factory designers seek it to minimize variations in indoor climate. In doing so they eliminate the element of life which can breathe vitality into otherwise sterile places and breathe joy into the human heart.

The conflicting requirements of urban compression and sunlight can sometimes be resolved in section. Where street sunlight is not critical the space can be narrowed, concentrating sunlight on building faces. Where sun on buildings is a problem (for glare and air-conditioning load — which says something about these sorts of buildings!) they can be turned so as to maximize sunlight to open spaces. Out of these simple functional responses to need-of-sun, a language of compressed and contrasting open spaces can emerge with sun-avenues and pathways, twisting according to cast shadow, matched with peak use times, much as one chooses the placing of windows to seek sun and light effects within rooms, especially at use times. Yet so many gables lack windows!

Shops that depend upon impulse purchases tend to be on the sunny side of the street. Banks, cinemas and places that are less dependent upon mood during the day tend to

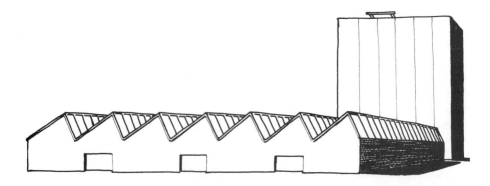

be on the shady side. The coincidence of the economic significance and soul-uplifting effects of sunlight gives support to sun- and shade-sensitive design. Street trade grows up around sunny steps as never it does in shady, draughty ravines. In California, sunlight rights (for economic purposes — solar collectors — not human souls) are protected by law!

Cities are concentrations of people together with their buildings, activities, transport and so on. They concentrate problems, often highlighting conflicting requirements. Just as spaciousness can be achieved in many ways, dimensional size being only one, many of the apparently irreconcilable polarities can be found to actually support each other if we look at things a different way. If we think of cities as *places* for *people*, urban problems no longer appear to be the inevitable result of urban civilization but the failure to grasp what people and place really need.

If we listen carefully to the needs of the human soul, differentially weighted and intensified by the urban situation, and to the needs of the spirit of a growing place, many opportunities begin to show themselves. Approaches to design can emerge in which nurture for the human spirit and economic benefit are coincident. We take for granted their opposite tendencies, yet they are, at heart, but different facets of the spirit of civilization. Commerce, after all, is the exchange of needed goods; civilization is the civilizing of commercial, social and cultural interaction.

In cities the enhanced opportunity arising out of multiple interactions is rich ground for entrepreneurial initiative. Though it need not be, most of this is so destructive of fine-grained textures, of organic development, of social patterns and of human freedoms. But cities depend upon life: they are (or should be) concentrations of civilization. And civilization is alive, something which is ever growing. If too conservative it can be claustrophobic.

All those great polarities of which cities are built make for an intensity of life which is both stressful and stimulating. Left to its own mechanisms — of which market forces and private defensive reaction are the predominant — the stresses will be soothed and the stimulation intensified, but only for the well-off and in ways which undermine civilization.

Unless we can develop opportunities for the life-renewing forces, forms and rhythms of nature to breathe into cities, as is done in the countryside, cities will be dominated by lifeless mechanical products. Unless we design for soul-nurture as a first priority, tranquillizers, vicarious experience and worse will take their toll on human freedom.

This isn't a matter of designing buildings for certain uses then making them look nice, but of doing it the other way around. It means starting by asking what sort of qualities — for *all* the senses — are appropriate to this situation, use or personal state. It means starting by listening, by thinking not of privately-delineated buildings but of communally-experienced place.

13

Building for Tomorrow

Most of us spend most of our time in, near or influenced by built surroundings. We spend our lives in what were once the thoughts of architects.[1] Our thoughts, our designs, make the world of tomorrow. If we think about it, this is a terrible responsibility; if we don't, there is no reason to suppose that it will be any less terrible a disaster than the effects of the recent past. In barely three decades architecture has destroyed cities and large areas of countryside all over the world. Pollution and environmental damage due to architectural decisions threaten even global ecology.

The ozone depletion crisis, for example, is to a significant extent due to chlorofluorocarbons used in air conditioning and foamed insulation. Much nuclear-produced electricity heats buildings (especially by night-storage heating).[2] Some materials, such as aluminium, require a lot of electrical energy to produce. Some, such as plastics, cause significant pollution in both manufacture and when their use life is over. There are other ways of heating and of cooling buildings. There are high and low pollution-price materials to choose from. Which systems and products we choose are architectural decisions.

Architecture has effects on place, on life-supporting ecology, on the spirit of the world we live in — it also affects people. In recent years Britain has witnessed a government committed to changing society by changes in communal responsibilities and economic structures. In a

[1] I use the word 'architect' throughout in its common sense meaning. Legally in Britain it means only architects registered by the Architects Registration Council UK (ARCUK). Most of us however live in surroundings not designed by ARCUK members nor perhaps well designed, but designed none the less.

[2] Nuclear power stations are best suited to a regular output. Peaks of demand can be topped up by conventional stations with a more flexible response. What do you do with nuclear-produced electricity when nearly everyone is in bed? Night-storage heating and attractively cheap tariffs to encourage off-peak usage are the solution.

short few years the human effects have become markedly visible, Major changes in our built environment over barely half a life-span have had at least as marked effects.

Before suburban box-land, urban filing cabinets and grab-for-yourself shopping, people lived differently: there were different unspoken values in society. It is not that everything is worse today — far from it — but one thing certainly is. Architecture is more finely geared to profitability then ever before. If it isn't, why are the professional journals full of these values? Even for those who choose not to work like this, architects have the power to make money for their clients. Indeed, some say that buildings are commissioned by people who want to make money out of them and the architect's role is to service this need. We may not all share this approach but it underlies the majority of the decisions which shape our world.

There is a school of thought which holds that encouraging profitability will establish a healthy economic base for run-down areas; 'quality of life' improvements follow in due course. Others hold that the dynamic management necessary for this does not want to live in a low-status dump. The arts, therefore, need subsidy as part of a management recruitment package. Both approaches have proved successful (in their terms) but I can't help feeling that something is the wrong way round; for the pursuit of profit has destroyed cities, countryside and bio-regions. It has destroyed individuals and divided society.

Yet profit unpursued is but a natural consequence of commercial activities which serve a need in the community: the interchange of goods and services arising out of listening to the surrounding situation. With this approach we can build a physical environment based on listening to the spirit of places and the needs of the human spirit. It won't solve all the problems, but it is a step in the right direction. To people who say you can't afford to work in this way I can only answer that I can.

With conventional economic structures, any aesthetic involvement in making things costs more money. Many people cannot afford more than the lovelessly utilarian (so I am told, though my observations of virtually all manufactured products is that they try to look good, usually to *look* better than they are). In fact, expense is much more a secondary issue than many people realize. In other parts of the world many things that we consider essential are unattainable luxuries. The crucial issue is, how long can we, as a society which hopes to remain civilized, survive if we give more value to use than beauty, to what we (privately and materially) can *get* out of things rather than what we (commonly and spiritually) can *give* through them?

In terms of spiritual nourishment deeper than the glossy cosmetic, much of our daily surroundings approach bankruptcy. The poor and less successful live in surroundings that are often aesthetic disasters; their values, sensitivities and dependence are pressured by these surroundings.

Partly as reaction, but mostly in the search for inner renewal from the deep well-springs of nature, some people seek solace in (relatively) wild environments. Wilderness, rough country, even more-or-less unpolluted and little-managed woodland, downland, heath and waterside are essential for the de-stressing, re-rooting and life-renewing which all too rarely can be found in most people's daily surroundings. None the less we can go to attractive 'natural' places — parks, woodland, moorland — and yet somehow not feel nourished. These are landscapes to look at and photograph but not to breathe in; they are landscapes in which we cannot feel a living spiritual presence. We can go to others where this life is very strong. You don't need to believe in fairies to experience this (although if you experience it strongly it may become hard to deny their existence!)

These are places which give us strength and renewal. Why? And what is it we experience there?

At the most material level we can observe perhaps that the air quality allows lichens to grow or that human activities (such as management by grazing animals) are in a harmonious balance with nature and the uniqueness of the place. Inevitably in such a place the ecology — be it lush or semi-arctic — is rich. It has so many biological pathways and cycles that you cannot make any one simple diagram. This gives it resilience and health. The elusive ambiguities of its multi-

I challenge any reader to find an ecologically healthy place which, however undramatic, is not also beautiful — and by beautiful I mean nourishing to the spirit. In the same way places which are ecologically one-sided tend to be one-sided in what they can offer in spiritual nutrition. For too long townscape has been dominated by the accidental consequence of compartmentalized accommodations of material needs. Can we as shapers of the built environment offer as wide and symbiotic a range of spirit nourishment as can healthy landscapes? Can we at least try to?

track systems make the whole place seem a living being.

These sort of places are food for our spirit; we can meaningfully use the word 'beauty', even where photographically they may be a bit uneventful. Aesthetics — a spiritual description — and ecological stability — a material one — are inseparable.

I sometimes have the experience that the weather is an outer picture of how I feel inside. At first sight this is a ridiculous idea and just goes to prove that I, and others who have this experience, are psychologically unhinged! But it is true that living weather has within it many moods; a wind can be both fierce and cleansing, sunlight both relaxing and life-stirring at the same moment; the clouds are never fixed; the weather is always in a living state of change.

Somewhere within these many, simultaneous, elusively indefinable moods are the moods that we need: moods that are outer pictures of our inner soul life, moods that give the balance that we need. In nature, even developed or disturbed by man (as it is *everywhere* in the world) we can find these moods. We can also try to provide as wide a choice of mood in the man-made environments we build. It is after all in these that 90 per cent of us spend 90 per cent of our time.

Before even we can start to think about places to nourish the soul we have to be emphatic that places are for people. This may sound obvious, but the fact is that most places are, to a large extent, the accidental result of collections of buildings, each conceived as a separate object.

Even the spaces within these buildings are often designed to provide for people as quantitative statistics to be packaged efficiently and lovelessly.

I was taught that planning starts with a 'bubble diagram' of relationships between different spaces; with no quality attached to the linking lines which so desperately need to in adverbs. Diagramatically, a lift is a perfect way to convey people, to link bubbles on the diagram. But if we think of the pleasure, sociability, experience progression and preparatory thresholds of the journey, we might choose a sloping, winding passageway opening to many views, passing and coloured by many events. Reading the diagram as adverbs rather than mere relationships between nouns would, therefore, lead to entirely different planning.

Starting by organizing diagrams while leaving qualities to be added subsequently is like painting by colouring-in drawings. Paintings live in colour — colour which is concentrated, enhanced and modified in its effects by shape and boundary. Likewise environments live in sensory quality — the effects of which are organized to concentrate particular moods. Diagrams play an important part in organizing the architect's thoughts, but they are only a starting point, the point of understanding what, arising out of the most practical requirements, are the important mood themes and how they can be brought together in terms of spatial relationship. From then on these moods and their relationships lead the design until what was once a

Entry arch to woodland path

Welcoming asymetrical entry gesture

Sunlit-protected play yard (brick paving)

Dark, low, turning 'portal passage' (tile floor)

Bright place to stop and remove boots (tile floor)

If sequential experiences are seen not merely as the consequences of a diagram, but as meaningful adverbs, they may well become the organizing features of a design. Here in the entry to a kindergarten it is important that small children leave the restrictive stress and abstract visual-only experiences of their car journeys to school behind them.

Turn of direction to a spacious room, now at tree-top level and with a wooden floor which sounds quite different from the previous brick or tile flooring

diagram has disappeared and been reborn in terms of experience sequences.

I was also taught that architects solve problems; design proposals were referred to as 'solutions'. People can be offended if you refer to their home as a problem, or a solution — to them it is a living being, rich in multiple functions both spiritual and practical, mostly inseparable.

The simplistic categorical nature of diagram thinking and problem solving tends to push the qualitative aspects into second place at least. It leads to 'covered way', 'corridor' and 'route' instead of cloister, passageway and alley. Diagram thinking can extend into adjectival descriptions like 'secluded study corner', but unless these key words 'secluded', 'study' and 'corner' can be brought alive there is a risk that the building will not rise above the level of a built diagram. I have been in many buildings where I know exactly how the architect intended me to act.

The diagram is not something to build. It stands for something richer, just as the written script for an actor is only the stepping-stone to that moment when suddenly great spiritual truths flow through his whole being, transforming his words from repetition to something to touch every heart that hears.

Many of those buildings that form our world, however, do not even rise above their allegiances to dead material — ease of industrial manufacture, material durability or monetary savings at the expense of life-supporting construction and design. Their qualitative characteristics are life-suppressing and their physical, biological and spiritual effect on places and on people are damaging.

These sort of buildings have set a trend in simplifying architecture down to that which is photographable. As such they appeal to other photographic-conscious architects, establishing an incestuous cycle which still has an enormous influence on the profession. It is at the price of all those other qualities that both arise from, and nurture, the realm of life.

I have discussed forms and qualities that are full of life. Yet these can be used in many ways. When science-fiction illustrators picture the future it is in an architecture of curvilinear towers and pinnacles, interwoven with fluid-formed trackways, often in shimmering materials. These astro-cities are set within harsh dead deserts, depending totally upon mechanical life-support. They themselves have no meaningful root in place, time or living processes; instead the forms

Photography can focus on architecture or its occupants. In fact, our attention is normally on the latter but architecture is setting the background mood. For everyone except architects the issue is not how noteworthy the architecture is, but what mood it sets.

of their architecture imitate the forms of living things. They are cities built on illusion and fantasy.

Such environments are not so far away. In advertising, children's toys, entertainment, restaurant and shop interiors and even in building forms, some of these powerful, deceptive, outer-image forms are already around us.

This fantasy has great appeal for it is the complete opposite of the world of faceless, organized, dead, mineral objects that can be found in every city all over the world. Typically, buildings for bureaucracies are anonymous, gridded boxes; bank headquarters rectangular, patterned towers lifelessly dominating the streetscape at their feet.

These sort of buildings — the fantastic and the rigid — are pic-

tures, with powerful soul effect, of inhuman polarities within society. One pole feeds all that is materialistic and so rationally organized that the ambiguous, unpredictable and spontaneous is suppressed. At the other extreme lies personal, emotional and physical indulgence, the cultivation of desires in place of responsibilities. A lot of money is made catering for and reinforcing this tendency. The lifeless realm, on the other hand, controls a lot of money and with it a lot of people's lives.

These forces are very much in evidence around us. To observe them we need only to step back from what we take for granted. They are forces which seek to diminish individual moral responsibility and inner freedom — in other words, that which stands at the core of the human being, distinguishing humans from animals. Some would call these forces diabolical and point to their manifestation in the arms race or narcotics trade — *other* people do *those* sorts of things. But whenever architecture seeks to influence people for gain or gives its allegiance to things rather than people or spirit of place, it is servant to these same masters.

Not just to counter these forces but also to bring something health-giving to the natural and human environment we need to bring together the organizing and the life-filled, the rational and the feeling, the straight and the curved, the substantial and the transitory, matter and light, in a different way. We need to build buildings and places of life-renewing, soul-nurturing, spirit-strengthening qualities. Soul can only be given by souls — not by computer systems or industrial might. These have their place as aids to fulfilling our intentions; too often however they have a shaping influence upon these intentions, leaving no room for living processes of design, construction, use and maturation. It is living processes that bring things to life.

This may sound like impractical nonsense, the opposite of the way things are normally done but I am convinced it is the only *sensible* way. Some people say that the approach set out in this book, and upon which my own architectural work is founded, stands the conventional world on its head, others that it is just ordinary common sense. I hope it is the latter — though if it is why don't we see more evidence of this common sense in our surroundings? All that I am trying to do is to stand back and recognize the essence of the situation, the essence of what is happening.

The course of every human life is uniquely individual, yet together we share certain biographical patterns. We develop not only physically but in early life towards becoming self-directed individuals and thereafter in our individual development. Outwardly we meet both stress and stimulation, obstacle and opportunity from environment and society. Inwardly we travel a path of transformation of lower egocentric and bodily-bound forces into higher forms which are more gift orientated, more spiritual. People make this journey with widely varying motivation, persistence, speed and success.

The earth itself is also on a journey of transformation. Man's deeds make more or less room for the health-giving forces of nature and cosmos to work. There are many, many transformations for the worse but also many for the better, potent even if less conspicuous.

That is the situation. This is the context in which our architectural deeds must be placed. Will they aid the fragile upward growth process or the powerful downward one?

Yet even after making conscious decisions to work for growth and freedom in place of ossification and enslavement (for love rather than for power) we need to know *how* to act.

Any action can be raised into an artistic deed, any experience to an artistic experience — and this underlies the dilemma in contemporary fine art. When does the work of art *require* observers to change their inner state so as to be able to experience what others consider to be banal as art? Or when is the *experience* of what is perhaps outwardly common place none the less so moving that it has the effect of an artistic experience without demanding to be called art?

The former category predominates in the commercial galleries — like the emperor's clothes they are after all well suited to status purchasers. The latter brings a raising, civilizing, enabling, healing influence into society.

It is easy just to state our intentions and then prove that we have achieved them. I am not immune from this, which is why I ask you to consider the contents of this book dispassionately, neither supporting nor objecting to what is potentially emotive material. Only by observing things in this way, without the fog of personal reaction, attempting to penetrate to the true essence of their being, can we hope to come to any objectively meaningful assessment. We need to look at the architectural vocabulary — in its widest sense — in this way. When we really observe what, for example, a hard smooth surface, mid-morning sunlight, acoustic absorbent materials and so on really *do*, we can get a sense of what they really do to living spirits — to places, to people, to ourselves.

Just as our inner development steers and is steered by our biography, we shape and are shaped by our environment. This cyclical process is so indissolubly bound that we cannot step outside it to shape or be shaped differently without conscious action. It is this step that this book is concerned with.

The issues are universal, but any applications involve, of course, individual situations and individual interpreters. I have given examples of how I go about doing things, examples which I hope make clear that there are no serious difficulties in working in this sort of way. Most categorically, however, I wish to avoid presenting a series of answers. There are no answers except those for which the seeds lie in every question; and every question is unique as it arises afresh in every new situation. Rather, my hope is to colour the whole way of going about things. It is like learning to speak, something to develop, cultivate, sharpen — but when you use it, to forget. This colour is, I hope, something to become part of our beings. For all of us it is not something fixed and final but something growing.

What I write is not novel; I write the obvious. It is my belief that we all already know it — and the test of my ability to transcend the limitations of my personal viewpoint is whether you recognize truth in what I describe or regard it as a lot of nonsense. We all know it, but somewhere the significance of *what* we know has slipped through the cracks in the floor. When I think for a moment of this significance, I am horrified by how little account is taken of it. I am horrified by the disastrous effects of the world we are building today.

My plea is that the obvious is taken seriously. If it isn't, we will be known as the generation of destroyers — destroyers of places, of ecological stability and of the *human* in human beings. If it is, we can start to build an architecture of healing, to build places of the soul.

This book is only about the start.

List of Photographs

Index